Introduction

The aim of the *Primary Mathematics* curriculum is to allow students to develop their ability in mathematical problem solving. This includes using and applying mathematics in practical, real-life situations as well as within the discipline of mathematics itself. Therefore, the curriculum covers a wide range of situations from routine problems to problems in unfamiliar contexts to open-ended investigations that make use of relevant mathematical concepts.

An important feature of learning mathematics with this curriculum is the use of a concrete introduction to the concept, followed by a pictorial representation, followed by the abstract symbols. The textbook supplies the pictorial and abstract aspects of this progression. You, as the teacher, should supply the concrete introduction. For some students a concrete illustration is more important than for other students.

This guide includes the following :

- **Scheme of Work**: A table with a suggested weekly schedule, the objectives for each lesson, and corresponding pages from the textbook, workbook, and guide.

- **Manipulatives**: A list of manipulatives used in this guide

- **Objectives**: Lists objectives for each chapter.

- **Notes**: Explains what students should know before starting the chapter, the concepts that will be covered in the chapter, and how they fit in with the program as a whole.

- **Material**: Lists suggested manipulatives that can be used in presenting the concepts in each chapter.

- **Activity**: Provides teaching activities for introducing a concept concretely or for following up on a concept to clarify or extend it so that students will be more successful with independent practice.

- **Discussion**: Indicates the opening pages of the chapter and tasks in the textbook that should be discussed with the student. A scripted discussion is not provided. You should follow the material in the textbook, but any additional pertinent points that should be included in the discussion are given in this guide.

- **Practice**: Indicates which tasks in the textbooks students could do as guided practice or as an assessment for you to see if they understood the concepts covered in the teaching activity or the discussions.

- **Workbook**: Gives the workbook exercise that should be done after the lesson.

- **Reinforcement**: Lists additional activities that can be used if your student needs more practice or reinforcement of the concepts. This includes the exercises in the optional *Primary Mathematics Extra Practice* book.

- **Games**: Provides optional simple games that can be used to practice skills.

- **Enrichment**: Provides optional activities that can be used to further explore the concepts or to provide some extra challenge.

- **Tests:** Refers to the appropriate tests in the *Primary Mathematics Tests* book.

- **Answers:** Provides answers to all the textbook tasks and workbook problems, and many fully worked solutions. Answers to textbook tasks are proved within the lesson, and answers to workbook exercises for the chapter are located at the end of each chapter in the guide.

- **Mental Math**: Problems for more practice with mental math strategies.

- **Appendix**: Pages containing drawings and charts that can be copied and used in the lessons.

The textbook and workbook both contain a review for every unit. You can use these in any way beneficial to your student. For students who benefit from a more continuous review, you can assign three problems a day or so from one of the practices or reviews. Or, you can use the reviews to assess any misunderstanding before administering a test. The reviews, particularly in the textbook, do sometimes carry the concepts a little further. They are cumulative, and so allow you to refresh your student's memory or understanding on a topic that was covered earlier in the year. In addition, there are supplemental books for *Extra Practice* and *Tests*. In the test book, there are two tests for each section. The second test is multiple choice. There is also a set of cumulative tests at the end of each unit. You do not need to use both tests. If you use only one test, you can save the other for review or practice later on. You can even use the review in the workbook as a test and not get the test book at all. So there are plenty of choices for assessment, review, and practice.

The mental math exercises that go along with a particular chapter or lesson are listed as reinforcement in the lesson. They can be used in a variety of ways. It is not expected that you use all the mental math exercises listed for a lesson on the day of the lesson. In some review lessons, they can be used for the independent work, since there is not always a workbook exercise. You can have your student do one mental math exercise a day, repeating some of them, at the start of the lesson or as part of the independent work. You can have your student do a 1-minute "sprint" at the start of each lesson using one mental math exercise for several days to see if the student can get more of the problems done each successive day. You can use the mental math exercises as a guide for creating additional "drill" exercises.

This "Scheme of Work" on the next few pages is a suggested weekly schedule to help you keep on track for finishing the textbook in about 18 weeks. No one schedule or curriculum can meet the needs of all students equally. For some chapters, your child may be able to do the work more quickly, and for others more slowly. Take the time your student needs on each topic and each lesson. For students with a good background, each lesson in this guide will probably take a day. For others, some lessons which include a review of previously covered concepts may take more than a day. There are also optional lessons that are entirely review and a few optional mental math lessons. Use the reinforcement or extension activities at your discretion. You may want to include extra days to play some of the games or to allow time for more practice, such as in memorizing the multiplication and division facts.

Scheme of Work

Textbook: *Primary Mathematics Textbook* 2B, Standards Edition
Workbook: *Primary Mathematics Workbook* 2B, Standards Edition
Guide: *Primary Mathematics Home Instructor's Guide* 2B, Standards Edition (this book)
Extra Practice: *Primary Mathematics Extra Practice* 2, Standards Edition
Tests: *Primary Mathematics Tests* 2B, Standards Edition

Week		Objectives	Text book	Work book	Guide
		Unit 7: Addition and Subtraction			
		Chapter 1: Finding the Missing Number			1-2
1	1	Find a missing number.	8-11	7-8	3-4
		Extra Practice, Unit 7, Exercise 1, pp. 103-104			
	2	Make 100.	11-12	9-10	5
	3	Practice.	13		6
		Tests, Unit 7, Chapter 1, A and B, pp. 1-4			
		Answers for unit 7 workbook exercises 1-2			7
		Chapter 2: Methods for Mental Addition			8-9
	1	Review: Add ones or tens to a 2-digit number.	15	11-12	10
	2	Review: Add ones, tens, or hundreds.	14-15	13-14	11
2	3	Review: Add two-digit numbers mentally.	15	15	12
	4	Add a number close to 100 to a 2-digit number.	16	16	13
	5	Add a number close to 100 to a 3-digit number.	16	17	14
		Extra Practice, Unit 7, Exercise 2, pp. 105-106			
		Tests, Unit 7, Chapter 2, A and B, pp. 5-8			
		Answers for unit 7 workbook exercises 3-7			15
		Chapter 3: Methods for Mental Subtraction			16-17
	1	Review: Subtract ones or tens from a 2-digit number.	18	18-19	18
	2	Review: Subtract ones, tens, or hundreds.	17-18	20-21	19
3	3	Review: Subtract two-digit numbers mentally.	18	22-23	20
	4	Subtract a number close to 100 from hundreds.	19	24	21
	5	Subtract a number close to 100 from a 3-digit number.	19	25	22
	6	Practice.	20		23
		Extra Practice, Unit 7, Exercise 3, pp. 107-110			
		Tests, Unit 7, Chapter 3, A and B, pp. 9-12			
		Answers for unit 7 workbook exercises 8-12			24
4		**Review 7**	21-23	26-29	25
		Tests, Units 1-7, Cumulative, A and B, pp. 13-16			
		Answers for unit 7 workbook review 8			26

	Week		Objectives	Text book	Work book	Guide
			Unit 8: Multiplication and Division			
			Chapter 1: Multiplying and Dividing by 4			27
		1	Count by 4's to multiply 4.	24-25	30-32	28
		2	Build multiplication tables for 4.	26-27	33-35	29
		3	Memorize multiplication facts for 4.			30
		4	Solve word problems.		36-39	31
22	5	5	Relate division by 4 to multiplication by 4.	27-28		32
		6	Memorize division facts for 4.			33
		7	Solve word problems and practice.	28-29	40-42	34
			Extra Practice, Unit 8, Exercise 1, pp. 115-118			
			Tests, Unit 8, Chapter 1, A and B, pp. 21-24			
			Answers for unit 8 workbook exercises 1-4			35
			Chapter 2: Multiplying and Dividing by 5			36
		1	Count by 5's to multiply by 5.	30-32	43	37
		2	Learn multiplication facts for 5, solve word problems.	32	44-45	38
23	6	3	Divide by 5.	32	46-47	39
		4	Solve word problems and practice.	33	48	40
			Extra Practice, Unit 8, Exercise 2, pp. 119-120			
			Tests, Unit 8, Chapter 2, A and B, pp. 25-28			
			Answers for unit 8 workbook exercises 5-7			41
			Chapter 3: Multiplying and Dividing by 10			42
		1	Multiply by 10.	34-35, 36	49-51	43
		2	Divide by 10.	35, 36	52-54	44
		3	Practice.	36-37		45
			Extra Practice, Unit 8, Exercise 3, pp. 121-122			
			Tests, Unit 8, Chapter 3, A and B, pp. 29-32			
			Answers for unit 8 workbook exercises 8-9			46
			Chapter 4: Division with Remainder			47
24	7	1	Find the remainder in division.	38-40	55-56	48
		2	Practice.	41		49
			Extra Practice, Unit 8, Exercise 4, pp. 123-124			
			Tests, Unit 8, Chapter 4, A and B, pp. 33-38			
			Review 8	42-43	57-60	50
			Tests, Units 1-8, Cumulative, A and B, pp. 39-45			
			Answers for unit 8 workbook exercise 10			51
			Answers for unit 8 workbook review 9			51

Week		Objectives	Text book	Work book	Guide
Unit 9: Money					
		Chapter 1: Dollars and Cents			52
8	1	Write money in dollars and cents.	44-46	61-63	53
	2	Read and write money amounts.	46	64-67	54
	3	Change from one denomination to another.	46-47	68-69	55-56
		Extra Practice, Unit 9, Exercise 1, pp. 129-132			
	4	Make change for $1.	48	70	57
	5	Make change for $5 and $10.	48	71	58
9	6	Solve word problems.	49	72-74	59
		Tests, Unit 9, Chapter 1, A and B, pp. 47-50			
		Answers for unit 9 workbook exercises 1-5			60
		Chapter 2: Adding Money			61
	1	Add dollars and cents mentally.	50-51	75-76	62
	2	Add dollars and cents using the addition algorithm.	52	77	63
	3	Add dollars and cents when the cents is almost $1.	52	78	64
	4	Solve word problems.	53	79-80	65
		Extra Practice, Unit 9, Exercise 2, pp. 133-136			
		Tests, Unit 9, Chapter 2, A and B, pp. 51-54			
		Answers for unit 9 workbook exercises 6-10			66
		Chapter 3: Subtracting Money			67
10	1	Subtract dollars and cents mentally.	54-55	81-82	68
	2	Subtract dollars and cents using the subtraction algorithm.	56	83	69
	3	Subtract dollars and cents when the cents is almost $1.	56	84	70
	4	Solve word problems.	57	85-86	71
		Extra Practice, Unit 9, Exercise 3, pp. 137-138			
	5	Practice.	58-59		72
		Tests, Unit 9, Chapter 3, A and B, pp. 55-58			
		Answers for unit 9 workbook exercises 11-15			73
11		**Review 9**	60-61	87-91	74
		Tests, Units 1-9, Cumulative, A and B, pp. 59-65			
		Answers for unit 9 workbook review 10			75

Week		Objectives	Text book	Work book	Guide
		Chapter 3: Time Intervals			101
14	1	Find the time interval in minutes or hours.	80-83	122-125	102-103
	2	Find the end time from the start time and the time interval.	83	126-127	104
		Extra Practice, Unit 11, Exercise 3, pp. 161-162			
		Tests, Unit 11, Chapter 3, A and B, pp. 115-122			
		Answers for unit 11 workbook exercises 3-4			105
		Chapter 4: Other Units of Time			106
	1	Understand other units of time.	84-85	128	107
		Extra Practice, Unit 11, Exercise 4, pp. 163-164			
		Tests, Unit 11, Chapter 4, A and B, pp. 123-127			
		Review 11	86-87	129-133	108
		Tests, Units 1-11, Cumulative, A and B, pp. 129-140			
		Answers for unit 11 workbook exercise 5			109
		Answers for unit 11 workbook review 12			109
Unit 12: Capacity					
		Chapter 1: Comparing Capacity			110
15	1	Compare the capacity of containers.	88-89	134-138	111
		Extra Practice, Unit 12, Exercise 1, pp. 169-170			
		Tests, Unit 12, Chapter 1, A and B, pp. 141-146			
		Chapter 2: Liters			112
	1	Measure capacity in liters.	90-94	139-140	113
		Extra Practice, Unit 12, Exercise 2, pp. 171-172			
		Tests, Unit 12, Chapter 2, A and B, pp. 147-152			
		Chapter 3: Gallons, Quarts, Pints and Cups			114
	1	Measure capacity in U.S. customary units.	95-98	141-142	115
		Extra Practice, Unit 12, Exercise 3, pp. 173-176			
		Tests, Unit 12, Chapter 3, A and B, pp. 153-158			
		Answers for unit 12 workbook exercises 1-3			116
		Review 12	99-100	143-147	117
		Tests, Units 1-12, Cumulative, A and B, pp. 159-170			
		Answers for unit 12 workbook review 13			118

Manipulatives

Whiteboard and Dry-Erase Markers
A whiteboard that can be held is useful in doing lessons while sitting at the table (or on the couch). Students can work problems given during the lessons on their own personal boards.

Multilink cubes
These are cubes that can be linked together on all 6 sides. Ten of them can be connected to form tens so that you have tens and ones. You can use Legos™ or anything else that can be grouped into tens, but the multilink cubes will be useful when the students get to volume problems in *Primary Mathematics* 4.

Counters
Round counters are easy to use and pick up, but any type of counter will work.

Base-10 set
A set usually has 100 unit-cubes, 10 or more ten-rods, 10 hundred-flats, and 1 thousand-block.

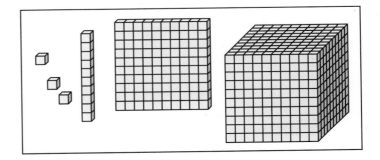

Place-value discs
These are round discs with 1, 10, 100, or 1000 written on them. You can label round counters using a permanent marker. You need 20 of each kind.

Thousands	Hundreds	Tens	Ones
1000 1000	100 100 100	10 10 10 10 10 10	1 1 1 1

Place-value chart
You can draw a simple one on paper and put it along with some cardboard backing into a sheet projector. That way, your student can use dry-erase markers to write numbers or draw number discs on the chart. Or just draw one whiteboard as needed during the lesson. It should be large enough to use with number discs or base-10 blocks.

Hundred-chart
A number chart from 1-100. Laminated ones are nice since you can use a dry-erase marker, or you can copy the one in the appendix.

Number cubes
Cubes that you can label and throw, like dice. You can use regular dice and label them with masking tape, or buy cubes and labels.

Number cards
Cards with 0, 1, 2, 3, 4, 5, 6, 7, 8, 9, or 10 written on them. You need four sets for games. Sometimes you will need some extra 0's. You can use playing cards with face cards removed. Use one set of face cards for 0, whiting out the J, Q, or K and replacing it with a 0. You can white out the Ace and replace it with a 1. Or create a deck of number cards from index cards or blank playing cards. You can also use an Uno™ deck with skips, wilds, and reverses removed.

Coins and Bills
Use play or real money. $20, $10, $5, and $1 bills, quarters, dimes, nickels, and pennies. The textbook does have some pictures of half-dollar coins, but those are not as common and if you don't have any you can just point them out in the text or look at images on the internet.

Store cards
Pictures of items with costs from $0.01 to $5.00. Cut pictures from magazines, newspaper ads, coupons, etc., glue to index cards, and write a cost, using decimal notation, e.g. $4.60. About half should be costs below $1 and another half costs between $1 and $5. To facilitate mental math computation, have the cents in multiples of 5.

Fraction bars and circles
You can copy the ones in the appendix or use commercially available fraction bars and circles.

Clock
One or two face clocks with geared hands. A large demonstration clock is nice and your student can use it too, but geared mini-clocks are sufficient.

Measuring cups
Quart and pint or cup measuring cups for liquid measurements. The quart measuring cup should also be marked in liters.

Solid shapes
Cubes, prisms, rectangular prisms, pyramids, cylinders, cones, and spheres. Preferably of similar sizes that the student can hold, examine, and compare.

Supplements

The textbook and workbook provide the essence of the math curriculum. Some students profit by additional practice or more review. Others profit by more challenging problems. There are several supplementary workbooks available at www.singaporemath.com. If you feel it is important that your student have a lot of drill in math facts, there are many websites that generate worksheets according to your specifications or provide on-line fact practice. Web sites come and go, but doing a search using the terms "math fact practice" will turn up many sites. Playing simple card games is another way to practice math facts.

Blank page

Unit 7 – Addition and Subtraction

Chapter 1 – Finding the Missing Number

Objectives

- Find the missing part in an addition or subtraction equation.
- Find the missing whole in an addition or subtraction equation.
- Use a count-on strategy to mentally subtract from 100.
- Use a missing-number strategy to mentally subtract from 100.

Notes

In *Primary Mathematics* 1A students learned to associate addition and subtraction with the part-whole concept and number bonds.

The number bond shown here uses ovals and has the parts on top. The important feature of the number bond is connecting the two parts to the whole, not whether the parts are on the top or bottom or a particular side, or whether there is an oval or a square or nothing around the numbers. At this level, drawing ovals or squares around the number helps to set them apart from each other.

When we know the parts, we use addition to find the whole.

When we know the whole and one of the parts, we use subtraction to find the missing part.

In *Primary Mathematics* 3, students will learn to draw bars rather than number bonds as a visual model. Some of the supplementary books available show bar models rather than number bonds even at the primary 2 level. However, at this age it can be difficult for students to draw the bar models neatly enough to interpret them. Number bonds are sufficient as a visual model for now.

In this chapter, students will use the part-whole model to find a missing number in addition and subtraction equations.

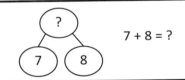

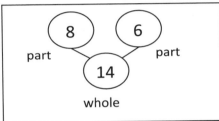

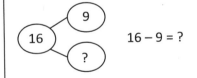

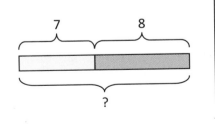

? + 9 = 17 missing a part	15 – ? = 8 missing a part	? – 7 = 4 missing a whole
↓	↓	↓
17 – 9 = ?	15 – 8 = ?	7 + 4 = ?

Students will learn two mental math strategies for finding the missing part when the total is 100, or for "making 100."

$$68 + \boxed{?} = 100$$

⇒ Count on, first by ones to the next ten and then by tens, or first by tens and then by ones:

$$68 \xrightarrow{+2} 70 \xrightarrow{+30} 100 \qquad \text{or} \qquad 68 \xrightarrow{+30} 98 \xrightarrow{+2} 100$$

⇒ Use the knowledge that 100 is 9 tens and 10 ones.

	6 tens	8 ones			6 tens	8 ones
+	? tens	? ones	⟶	+	3 tens	2 ones
	9 tens	10 ones			9 tens	10 ones

$$68 + \boxed{32} = 100$$

Your student may be able to use mental math strategies learned in earlier levels to add or subtract 1-digit, 2-digit, and some 3-digit numbers. Allow her to solve the problems mentally in this chapter when she can, or to rewrite the problems vertically and use the addition or subtraction algorithm when needed. Mental math strategies, other than subtracting from 100, will be reviewed more thoroughly in the next chapter, and new ones introduced.

When interpreting word problems in the practice at the end of this chapter, your student can use the part-whole concept to determine whether to add or subtract to find the answer. He should read the problem carefully, preferably out loud, and determine whether it gives two parts and is asking for a total or gives a total and one part and is asking for the missing part. He can draw a number bond and fill it in with information from the problem to help him determine what equation to use.

Some word problems involve comparison; that is, problems where we are told how much more or less one quantity is than another. In problems that ask for how much more or less one quantity is than another, a first step is to determine which is the larger quantity. The problem can be illustrated with a number bond or a simple drawing, if needed. For example, in problem 5 on p. 13 of the text, a cabbage weighs 324 g and a cucumber weighs 86 g less than the cabbage. The student is asked to find the weight of the cucumber. We can treat the larger quantity as the whole, the smaller quantity as one part and the difference as another part. Or your student can draw the items on a balance and label with quantities. This is better than making her draw a formal comparison model, which she will learn in order to solve 2-step problems in *Primary Mathematics 3*.

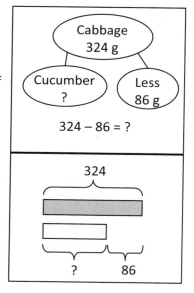

Material

♦ Counters
♦ 10 x 10 grids (Appendix p. a17)
♦ 2 number cubes, one labeled 1-6 and the other 4-9
♦ Mental Math 1-3 (Appendix)

(1) Find a missing number

Activity

Set out 14 counters or other small objects and cover up 6 of them with an index card so that there are 8 uncovered. You can make up a story, for example: "There are 14 children at a party. 8 of them are girls and the rest are boys. How many boys are there?" Write the equation with a missing addend: **8 + ___ = 14**. Your student will probably be able to tell you the missing number. Tell her that one part is 8, but we don't know the other part. Draw a number bond and tell her that since we need to find a missing part, we can write a subtraction equation. Write the subtraction problem **14 – 8 = ___**. We can use subtraction to find the answer to an addition problem with a missing part.

Now write the problem **14 – ___ = 8**. Tell your student that this problem tells him that there were 14 to start with, some were taken away, and 8 are left. Draw an empty number bond and ask him to fill it in to represent this problem. As in the previous example, 14 is the total, and 8 is a part. We want to find a missing part. So we can again use subtraction: 14 – 8 = ___.

Write the problem **___ – 8 = 14**. See if your student can make up a story for the problem. For example: "There were some children, 8 went away, and 14 were left." Ask if 14 is the total. It is not; the total is the number that is missing. Guide her in filling in a number bond and ask her what equation we can use to find the missing number. We need to add 8 and 14.

Discussion

Concept page 8

Tasks 1-2, pp. 9-11

For each of these, have your student identify the whole and determine if there is a missing part that needs to be found, or a total. Only the problem in 2(c) is missing a total; the rest are missing a part. Your student should be able to find the answers mentally, but If she has difficulty have her fill in a number bond for the problem and write the equation that can be used to find the missing number.

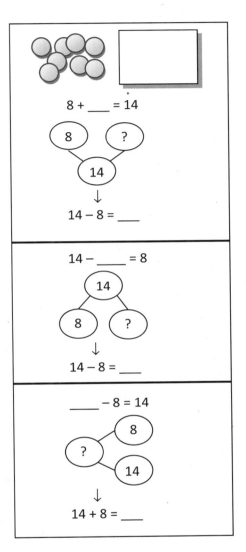

$8 + \underline{\quad} = 14$

$14 - 8 = \underline{\quad}$

$14 - \underline{\quad} = 8$

$14 - 8 = \underline{\quad}$

$\underline{\quad} - 8 = 14$

$14 + 8 = \underline{\quad}$

1. $20 - \boxed{} = 12 \longleftrightarrow 20 - 12 = \mathbf{8}$

2. (a) $14 + \boxed{} = 30 \longleftrightarrow 30 - 14 = \mathbf{16}$

   (b) $21 - \boxed{} = 5 \longleftrightarrow 21 - 5 = \mathbf{16}$

   (c) $\boxed{} - 6 = 18 \longleftrightarrow 18 + 6 = \mathbf{24}$

Practice

Task 3, p. 11

Workbook

Exercise 1, pp. 7-8
(Answers p. 7)

 Reinforcement

3. (a) $4 + \square = 13 \longleftrightarrow 13 - 4 = \mathbf{9}$ (b) $20 - \square = 13 \longleftrightarrow 20 - 13 = \mathbf{7}$

(c) $9 + \square = 39 \longleftrightarrow 39 - 9 = \mathbf{30}$ (d) $47 - \square = 19 \longleftrightarrow 47 - 19 = \mathbf{28}$

(e) $\square + 8 = 15 \longleftrightarrow 15 - 8 = \mathbf{7}$ (f) $\square - 7 = 10 \longleftrightarrow 10 + 7 = \mathbf{17}$

(c) $\square + 14 = 60 \longleftrightarrow 60 - 14 = \mathbf{46}$ (f) $\square - 16 = 40 \longleftrightarrow 40 + 16 = \mathbf{56}$

Extra Practice, Unit 7, Exercise 1, pp. 103-104

Give your student some additional problems to solve, particularly a few that cannot be easily answered mentally. You can use some of the these to review the addition and subtraction algorithm if needed.

$\Rightarrow$ $23 + \underline{\hspace{1cm}} = 37$ $\longleftrightarrow$ $37 - 23 = \underline{\hspace{1cm}}$ (14)

$\Rightarrow$ $\underline{\hspace{1cm}} + 42 = 60$ $\longleftrightarrow$ $60 - 42 = \underline{\hspace{1cm}}$ (18)

$\Rightarrow$ $76 - \underline{\hspace{1cm}} = 4$ $\longleftrightarrow$ $76 - 4 = \underline{\hspace{1cm}}$ (72)

$\Rightarrow$ $\underline{\hspace{1cm}} - 24 = 48$ $\longleftrightarrow$ $24 + 48 = \underline{\hspace{1cm}}$ (72)

$\Rightarrow$ $\underline{\hspace{1cm}} - 2 = 88$ $\longleftrightarrow$ $88 + 2 = \underline{\hspace{1cm}}$ (90)

$\Rightarrow$ $486 + \underline{\hspace{1cm}} = 925$ $\longleftrightarrow$ $925 - 486 = \underline{\hspace{1cm}}$ (439)

$$\begin{array}{r} \overset{8}{\cancel{9}}\;\overset{\overset{1}{1}}{\cancel{2}}{}^{1}5 \\ -\;\;4\;8\;6 \\ \hline 4\;3\;9 \end{array}$$

$\Rightarrow$ $\underline{\hspace{1cm}} - 256 = 474$ $\longleftrightarrow$ $474 + 256 = \underline{\hspace{1cm}}$ (730)

$$\begin{array}{r} 1\;\;1\;\;\; \\ 4\;7\;4 \\ +\;2\;5\;6 \\ \hline 7\;3\;0 \end{array}$$

Enrichment

Coded number puzzles can be used to find a missing addend and apply concepts of addition and subtraction. In them, digits are replaced by a symbol or letter. If the same symbol or letter appears more than once in the problem, it represents the same number. Number puzzles are common in problem solving books and you can make them up yourself. Here are a few:

$$\begin{array}{r} \&\;\;\&\;\;3 \\ +\;\;?\;\;2\;\;\& \\ \hline \%\;\;2\;\;?\;\;8 \end{array}$$

Since $3 + 5 = 8$, the & must be 5. Replace the other &'s with 5. The ? then must be 7. Replace those. The % must be 1.

$$\begin{array}{r} 2\;\;@\;\;9 \\ +\;\;\#\;\;\#\;\;4 \\ \hline @\;\;0\;\;\# \end{array}$$

The # must be 3. The 10 is renamed. Replace the #'s with 3, and write a 1 above the @. Now you have $1 + @ + 3 = 0$. Since it can't equal 0, it must equal 10. @ must therefore be 6.

$$\begin{array}{r} 3\;\;5\;\;A \\ -\;\;B\;\;C\;\;9 \\ \hline C\;\;C\;\;7 \end{array}$$

Since 7 is smaller than 9, the problem must require renaming. $\square - 9 = 7$ is the same as $9 + 7 = \square$. So $\square$ is 16, and A is 6. Cross out the 5 and replace with 4. The only possibility for C is 2. So B must be 1.

(2) Make 100

Activity

Write the problem **42 + ___ = 100** and have your student fill in a number bond, leaving a missing part.

Use a 10 by 10 array, such as the ones in the appendix on p. a17, or the reverse side of a hundred-board. Fill in or mark off 42 squares, as shown here. Use this to show your student visually that we can count on from 42 to 100, first by ones to get to the next ten, and then by tens to get to 100, or first by tens to get to a number between 90 and 100, and then by ones to get to 100.

Then point out that 100 is the same as 9 tens and 10 ones. We can think of 42 split into 40 and 2, and then find the missing numbers for 40 + ___ = 90 and 2 + ___ = 10. Those will be the tens and ones for the missing number in 42 + ___ = 100.

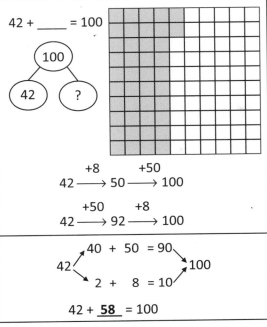

$$42 + \underline{\quad} = 100$$

$$42 \xrightarrow{+8} 50 \xrightarrow{+50} 100$$

$$42 \xrightarrow{+50} 92 \xrightarrow{+8} 100$$

$$42 \begin{array}{c} 40 + 50 = 90 \\ 2 + 8 = 10 \end{array} 100$$

$$42 + \underline{\textbf{58}} = 100$$

Since this missing number problem is the same as 100 − 42 = ___, we can use these strategies, counting on or making a 9 with the tens digit and a 10 with the ones digit, to subtract from 100.

Discussion

Tasks 4-5, pp. 11-12

Practice

Tasks 6-7, p. 12

Workbook

Exercise 2, pp. 9-10 (Answers p. 7)

Reinforcement

Mental Math 1-2

Enrichment

Show your student that we can use the same idea to mentally subtract from 1000. 1000 is the same as 9 hundreds, 9 tens, and 10 ones. So we can solve 1000 − 456, for example, with (900 − 400) + (90 − 50) + (10 − 6). Give her a few problems involving subtraction from 1000

⇒ 456 + ___ = 1000   (544)

⇒ 32 + ___ = 1000   (968)

⇒ 7 + ___ = 1000   (993)

⇒ 1000 − 405 = ___   (595)

⇒ 1000 − 63 = ___   (937)

4. 32

5. 53 + **47** = 100
   100 − 53 = **47**
   4 tens   **7** ones

6. (a) 66   (b) 24
   (c) 18   (d) 91

7. (a) 74   (b) 39   (c) 58
   (d) 4   (e) 98   (f) 92

$$456 \begin{array}{c} 400 + 500 = 900 \\ 50 + 40 = 90 \\ 6 + 4 = 10 \end{array} 1000$$

$$456 + \underline{\textbf{544}} = 1000$$

$$32 \begin{array}{c} 0 + 900 = 900 \\ 30 + 60 = 90 \\ 2 + 8 = 10 \end{array} 1000$$

$$32 + \underline{\textbf{968}} = 1000$$

(3) Practice

Practice

Practice A, p. 13

Your student can do this practice with you as a lesson or independently. You may want to review word problems with him using problems 2-5. Ask him pertinent questions, such as, "Does the problem give you a total amount? What is it?"

Your student is not required to use mental math to solve all the word problems in this and other practices. She should learn to determine when mental math is appropriate or when it might be better to use the addition or subtraction algorithm, rewriting the problem vertically. From the lessons in this chapter she should use mental math for #1(g,i-l) and #2, and if she has done earlier levels of *Primary Mathematics*, she should be able to do the rest of the problems in #1 mentally. She may prefer to use the standard algorithms on the word problems.

Reinforcement

Label two number cubes, one with 1-6 and the other with 4-9. Roll the two cubes. Allow your student to pick one to be the tens and the other to be ones. For example, if a 6 and a 2 is rolled the number can be 62 or 26. She must give the number that makes 100 with this number.

Mental Math 3

Test

Tests, Unit 7, Chapter 1, A and B, pp. 1-4

✳ Enrichment

Draw a picture of 5 squares as shown here. Write down a set of 5 numbers. 4 of them should be pairs that make 100, and the fifth a random number, for example, give him the numbers 45, 62, 23, 55, and 38. (45 and 55, 62 and 38 are pairs that make 100.) Ask your student to place the numbers in the squares so that they add up to the same number across and down. If he has trouble, tell him that there are two pairs that make 100 and see if he can find them. Help him see that they can go on the arms, and the fifth number in the middle. Repeat with other combinations.

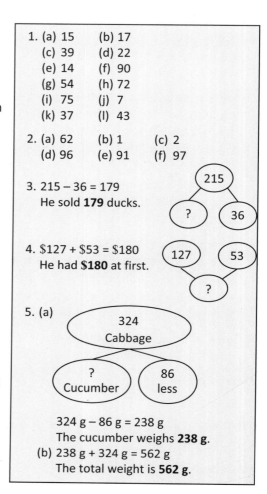

1. (a) 15   (b) 17
   (c) 39   (d) 22
   (e) 14   (f) 90
   (g) 54   (h) 72
   (i) 75   (j) 7
   (k) 37   (l) 43

2. (a) 62   (b) 1   (c) 2
   (d) 96   (e) 91   (f) 97

3. 215 − 36 = 179
   He sold **179** ducks.

4. $127 + $53 = $180
   He had **$180** at first.

5. (a)

324 g − 86 g = 238 g
The cucumber weighs **238 g**.
(b) 238 g + 324 g = 562 g
The total weight is **562 g**.

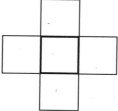

Fill in the squares with the numbers 45, 62, 23, 55, and 38 so that the sum of the numbers across is the same as the sum of the numbers down.

Workbook

Exercise 1, pp. 7-8

1. (a) 28 − 12 = **16**
   (b) 18 − 6 = **12**
   (c) 7 + 9 = **16**

2. U 88 − 68 = **20**   E 90 − 20 = **70**
   O 50 − 15 = **35**   T 58 − 18 = **40**
   E 80 − 35 = **45**   R 15 + 15 = **30**
   Y 91 − 16 = **75**   H 35 + 65 = **100**
           T 30 + 50 = **80**

   YOUR TEETH

Exercise 2, pp. 9-10

1. 10   70
   40   2
   95   85

2. (a) 1   (b) 5
   (c) 4   (d) 9
   (e) 20   (f) 65
   (g) 16   (h) 37
   (i) 58   (j) 42
   (k) 94   (l) 91

3. 80   90
   60   30

4. (a) 2   (b) 7
   (c) 15   (d) 73
   (e) 21   (f) 44
   (g) 78   (h) 66
   (i) 91   (j) 93
   (k) 99   (l) 96

Chapter 2 – Methods for Mental Addition

Objectives

♦ Add ones or tens to a 2-digit number.
♦ Add ones, tens, or hundreds to a 3-digit number.
♦ Add two 2-digit numbers.
♦ Add a number close to 100 to a 2-digit number.
♦ Add a number close to 100 to a 3-digit number.

Notes

In this chapter your student will review mental math strategies for addition learned in earlier levels of *Primary Mathematics* and will learn some new strategies. These strategies include the following:

• Add 1, 2, or 3 by counting on.

$$39 + 2 = 41 \qquad \text{Count on: } 40, 41$$
$$489 + 3 = 492 \qquad \text{Count on: } 490, 491, 492$$

• Add two 1-digit numbers whose sum is greater than 10 by making a 10. (This strategy is useful for students who know the addition and subtraction facts through 10, but have trouble memorizing some of the addition and subtraction facts through 20.)

$$7 + 5 = 10 + 2 = 12 \qquad\qquad 7 + 5 = 10 + 2 = 12$$
$$\overset{\displaystyle\wedge}{3 \quad 2} \qquad\qquad\qquad\qquad \overset{\displaystyle\wedge}{2 \quad 5}$$

• Add tens or hundreds using the same strategies used for adding ones.

$$293 + 20 = 313 \qquad \text{Count on } 303, 313$$
$$70 + 50 = 120 \qquad 7 \text{ ones} + 5 \text{ ones} = 12 \text{ ones} \rightarrow 7 \text{ tens} + 5 \text{ tens} = 12 \text{ tens}$$
$$700 + 500 = 1200 \quad 7 \text{ ones} + 5 \text{ ones} = 12 \text{ ones} \rightarrow 7 \text{ hundreds} + 5 \text{ hundreds} = 12 \text{ hundreds}$$

• Add tens by just adding the tens or by making the next hundred.

$$22 + 60 = 80 + 2 = 82 \qquad\qquad 480 + 50 = 500 + 30 = 530$$
$$\overset{\displaystyle\wedge}{2 \quad 20} \qquad\qquad\qquad\qquad \overset{\displaystyle\wedge}{20 \quad 30}$$

• Add a 1-digit number when there is no renaming by simply adding the ones together.

$$45 + 2 = 40 + 7 = 47 \qquad\qquad 645 + 2 = 640 + 7 = 647$$
$$\overset{\displaystyle\wedge}{40 \quad 5} \qquad\qquad\qquad\qquad \overset{\displaystyle\wedge}{640 \quad 5}$$

• Add a 1-digit number where adding the ones results in a number greater than 10

⇒ by making the next 10,

$$67 + 5 = 70 + 2 = 72 \qquad\qquad 467 + 5 = 470 + 2 = 472$$
$$\overset{\displaystyle\wedge}{3 \quad 2} \qquad\qquad\qquad\qquad \overset{\displaystyle\wedge}{3 \quad 2}$$

⇒ or by using basic addition facts.

$$67 + 5 = 60 + 12 = 72 \qquad\qquad 467 + 5 = 460 + 12 = 472$$
$$\overset{\displaystyle\wedge}{60 \quad 7} \qquad\qquad\qquad\qquad \overset{\displaystyle\wedge}{460 \quad 7}$$

- Add a 2-digit number to a 2-digit number

  ⇒ by adding the tens and then the ones, using the strategies already learned.

  $$48 + 36 = 48 + 30 + 6 = 78 + 6 = 84$$

- Add a number close to 100 (these strategies are new here)

  ⇒ by making 100,

  $$57 + 98 = 55 + 100 = 155$$
  $$\wedge$$
  $$55 \quad 2$$

  ⇒ or by first adding 100 and then subtracting the difference.

  $$57 + 98 = 57 + 100 - 2 = 157 - 2 = 155$$

Learning and using mental calculation strategies encourages flexibility in thinking about numbers and helps the student develop a strong number sense. Flexibility is a key here. Though your student will learn specific strategies, he should be encouraged to develop, utilize, and share his own strategies. Do not discourage any attempts to manipulate numbers. Do not require him to write down his steps, or explain his steps in writing. This negates the usefulness of mental math. Encourage your student to share his ideas and methods orally.

By now, your student should be able to easily make the next ten for a 2-digit number and add 2-digit numbers mentally. If she struggles with mental math, you could have her draw simple number bonds but be careful to not substitute another paper and pencil method for the standard algorithm at this stage, such as the one shown to the right where the student must use a "branching" method and write the tens to the outside and another branch to show how he is making a ten. This defeats the purpose of mental math and imposes inflexibility and arbitrary restrictions on mental math processes by requiring students to follow a specific procedure. It also bypasses the strategy emphasized in the *Primary Mathematics* curriculum of making the next ten. It would be better to go back and have her learn the concepts as they are taught in earlier levels of *Primary Mathematics* than to have her learn steps for a new paper and pencil method that only works with 2-digit numbers and bypasses the earlier foundation.

$$67 + 25$$
$$\wedge \qquad \wedge$$
$$60 \quad 7 \quad 5 \quad 20$$
$$\wedge$$
$$3 \quad 2$$
$$= 60 + 10 + 20 + 2 = 92$$

In other units where mental math is not specifically being taught, do not require your student to use mental calculations in problems where she is more comfortable using the addition algorithm. A problem written horizontally in the textbook does not mean it *has* to be solved mentally; it is just written horizontally to conserve space. She should determine when and where to try mental math in order to get an answer she is confident is correct.

Material

- Place-value discs
- 10 x 10 grids (Appendix p. a17)
- Mental Math 4-10 (Appendix)

(1) Review: Add ones or tens to a 2-digit number

Activity

Have your student solve the following problems mentally. Discuss possible strategies. Use place-value discs and/or number bonds to illustrate the possible strategies.

⇒ 7 + 5

Remind your student that we can find the answer by making a ten with 7 by adding 3 from the 5, leaving 2, so the answer is 10 and 2, or 12.

⇒ 32 + 5

Adding ones won't make another ten, so add the ones.

⇒ 37 + 5

Make the next ten, which is 40, by taking 3 from 5, leaving 2 ones, so the answer is 42. You can show this with place-value discs by moving over 3 ones from the second number.

Since we know there will be one more ten, we can write down or think 4 tens and then find the ones by remembering that 7 + 5 = 12, so the ones will be 2. You can show this with place-value discs by moving over all 5 ones and replacing 10 ones with a ten.

⇒ 70 + 50

Since 7 ones + 5 ones = 12 ones, 7 tens + 5 tens = 12 tens. Or, make the next 100, leaving 20.

⇒ 72 + 50

First add the tens in the same way as before, and then the 2 ones.

Practice

Task 1, p. 15

Workbook

Exercise 3, pp. 11-12 (Answers p. 15)

Reinforcement

Mental Math 4-5

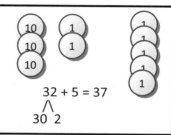

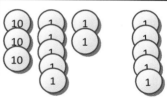

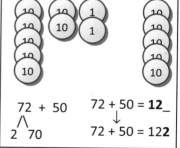

1. (a) 49   (b) 45   (c) 61
   (d) 50   (e) 90   (f) 140
   (g) 54   (h) 93   (i) 147

(2) Review: Add ones, tens, or hundreds

Discussion

Concept page 14

Point out that we add tens to tens, ones to ones, and hundreds to hundreds. Illustrate the problems on this page with actual place-value discs if necessary. Remind your student then when we change a place by a small number, we can count on: 356 + 20: 366, 376.

Activity

Have your student solve the following problems mentally. Discuss strategies. Use place-value discs and/or number bonds to help your student understand the strategies as needed.

⇒ 248 + 5

Make the next ten, using 2 from the 5, which leaves 3 ones, so the answer is 253.

⇒ 753 + 9

Make the next ten, taking 7 from 9 leaving 2 ones, so the answer is 762. Or, make a ten with the 9, taking 1 from the 3 ones of 573. Or, add ten and subtract 1, since 9 is one less than 10.

⇒ 456 + 2

The hundreds and tens don't change when we add ones, so just add the ones.

⇒ 498 + 5

Your student should recognize that this number is almost 500. Make the next hundred, using 2 from the 5, so that there are 3 ones in the answer.

⇒ 230 + 80

Add 23 tens + 8 tens in the same way that as for 23 + 8, and then add a 0 since the answer is tens, rather than ones.

⇒ 233 + 80

Add this in the same way as 230 + 80, and then write 3 for the ones, rather than 0.

Practice

Task 2, p. 15

Workbook

Exercise 4, pp. 13-14 (Answers p. 15)

Reinforcement

Mental Math 6-7

$248 + 5 = 250 + 3 = 253$

    ∧
   2 3

$753 + 9 = 760 + 2 = 762$

    ∧
   7 2

$753 + 9 = 752 + 10 = 762$

∧
2 1

$456 + 2 = 458$

   ∧
450 6

$498 + 5 = 500 + 3 = 503$

    ∧
   2 3

$230 + 80 = 300 + 10 = 310$

    ∧
  70 10

$230 + 80 = 300 + 10 = 310$

   ∧
220 10

$233 + 80 = 310 + 3 = 313$

∧   ∧
3 230 70 10

---

2. (a) 162   (b) 283   (c) 612
   (d) 380   (e) 305   (f) 214
   (g) 400   (h) 800   (i) 900
   (j) 456   (k) 904

(3) Review: Add two-digit numbers mentally

Activity

Have your student solve the following problems mentally. Discuss possible strategies. Use place-value discs and/or number bonds to illustrate the strategies.

⇒ 43 + 35

From the place-value discs, we can easily see that we can add these numbers by simply adding the tens together and the ones together.

⇒ 47 + 35

We can still add the tens and the ones separately, but there will be an extra ten. First add the tens, to get another two digit number, and then the ones, using the same strategies used to add a 1-digit number to a 2-digit number.

Rather than writing down an intermediate number, we can find each digit from left to right:

1. Find the tens. Look ahead to the ones – will adding the ones increase the tens? Yes, so add the tens and add one more ten. Write or think 8 for the tens.

2. Find the ones: 7 + 5 = 12, so the ones is 2. Write or think 2 for the ones. The answer is 82.

Practice

Tasks 3-4, p. 15

Workbook

Exercise 5, p. 15 (Answers p. 15)

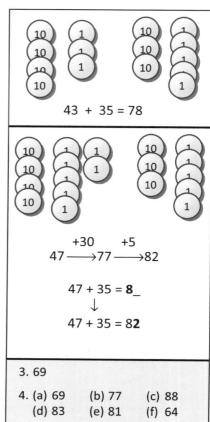

43 + 35 = 78

+30   +5
47 ——→77 ——→82

47 + 35 = **8**_
↓
47 + 35 = **82**

3. 69

4. (a) 69   (b) 77   (c) 88
   (d) 83   (e) 81   (f) 64

Reinforcement

Use 4 sets of number cards 1-9. Shuffle. Turn the first two over and ask your student to tell you if adding them will give a number greater than ten or less than ten without finding the actual answer. Repeat, turning over two cards at a time.

Mental Math 8

Game

Material: Two decks of playing cards with the face cards and tens removed. Separate the red and black cards into two separate decks.

Procedure: Shuffle and deal out the red and black cards, keeping them separate. Each player turns over two cards from each of their decks and forms two 2-digit numbers, with the red cards as tens, and adds the numbers. The player with the highest sum gets all the cards that have been turned over. If two players have the same sum, the player with the highest number on any single card gets all the cards. The player with the most cards after all the cards have been turned over wins.

Add a number close to 100 to a 2-digit number

Activity

Write the following problems and discuss strategies for solving them mentally, using number bonds and two 10 x 10 grids to illustrate the process.

⇒ 98 + 6

Make a hundred, taking 2 from the 6, leaving 4 ones.

⇒ 98 + 36

Make 100 here as well, by taking 2 from 36. This leaves 34, so the answer is 134.

Show the number 98 on one 10 x 10 grid and 36 on another. Remind your student that 98 + 36 = 36 + 98 if needed. Point out that 98 is almost 100.

Ask your student what would happen if we just added 100 to 36 to get 136. How does that compare to adding 98? It is 2 more, since 100 is 2 more than 98. So we could add 100 to 36, and then subtract 2 to get the correct answer.

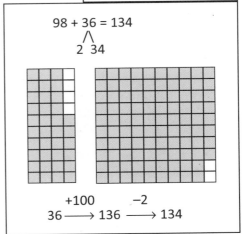

⇒ 99 + 42

Add 100 to 42 and subtract 1, since 99 is 1 less than 100. Think 100 more and then 1 less: 142, 141.

$$99 + 42 = 142 - 1$$
$$= 141$$

⇒ 69 + 97

Ask your student what we would need to subtract if we added 100 to 69 in order to get the correct answer. We can think 100 more and then 3 less: 169, 166

$$69 + 97 = 169 - 3$$
$$= 166$$

⇒ 51 + 98

Add 100, and subtract 2 by counting down. 151, 150, 149.

$$51 + 98 = 151 - 2$$
$$= 149$$

Practice

Tasks 5-7, p. 16

Workbook

Exercise 6, p. 16 (Answers p. 15)

Reinforcement

Mental Math 9

5. 103

6. (a) 101 (b) 108 (c) 103
   (d) 145 (e) 157 (f) 134

7. (a) 127 (b) 153 (c) 194
   (d) 155 (e) 184 (f) 197

(5) Add a number close to 100 to a 3-digit number

Activity

Write the following problems and discuss strategies for solving them mentally, using number bonds to illustrate the process.

⇒ 643 + 99

Make 100 with 99, so the answer is 100 + 642.

Ask your student to compare the answers for 643 + 99 and 643 + 100. When we add 100, we are adding 1 more than when we add 99, because 100 is 1 more than 99. So a quick way to add 99 is to add 100 and subtract 1; that is, count up by 100 and back by 1.

$$643 + 99 = 642 + 100 = 742$$

$$642 \quad 1$$

$$643 \xrightarrow{+100} 743 \xrightarrow{-1} 742$$

⇒ 384 + 98

Count up 100 and back 2.

$$384 + 98 = 484 - 2$$
$$= 482$$

⇒ 422 + 97

Count up 100 and back 3.

$$422 + 97 = 522 - 3$$
$$= 519$$

⇒ 98 + 301

Count up 100 from 301 and back 2.
Or, simply add 300 and then 1 to 98: 398 + 1.

$$98 + 301 = 401 - 2$$
$$= 399$$

⇒ 489 + 92

Add 100 and subtract 8.

$$489 + 92 = 589 - 8$$
$$= 581$$

Practice

Tasks 8-9, p. 16

Workbook

Exercise 7, p. 17 (Answers p. 15)

8. 336

9. (a) 355   (b) 406   (c) 751
   (d) 202   (e) 561   (f) 397

Reinforcement

Mental Math 10

Extra Practice, Unit 7, Exercise 2, pp. 105-106

Test

Tests, Unit 7, Chapter 2, A and B, pp. 5-8

Workbook

Exercise 3, pp. 11-12

1. (a) 2   (b) 60   (c) 38
   (d) 40   (e) 80   (f) 27

2. (a) 27   (b) 58
   (c) 69   (d) 48
   (e) 88   (f) 79
   (g) 34   (h) 71
   (i) 28   (j) 80
   (k) 40   (l) 64
   (m) 51   (n) 97
   (o) 63   (p) 54

3. (a) 60   (b) 60
   (c) 90   (d) 70
   (e) 120   (f) 130
   (g) 130   (h) 130
   (i) 120   (j) 150
   (k) 140   (l) 180

4. (a) 45   (b) 68
   (c) 96   (d) 87
   (e) 103   (f) 109
   (g) 102   (h) 109
   (i) 118   (j) 123
   (k) 107   (l) 122

Exercise 4, pp. 13-14

1. (a) 166   (b) 235
   (c) 409   (d) 407
   (e) 788   (f) 659

2. (a) 141   (b) 196
   (c) 362   (d) 415
   (e) 572   (f) 664
   (g) 743   (h) 298

3. (a) 260   (b) 549
   (c) 482   (d) 658
   (e) 765   (f) 375
   (g) 677   (h) 893

4. (a) 310   (b) 500
   (c) 728   (d) 615
   (e) 466   (f) 955
   (g) 845   (h) 780

5. (a) 400   (b) 800
   (c) 900   (d) 500
   (e) 800   (f) 700
   (g) 600   (h) 900

6. (a) 450   (b) 706
   (c) 675   (d) 909
   (e) 664   (f) 725
   (g) 915   (h) 935

Exercise 5, p. 15

1. (a) 70   (b) 4   (c) 89

2. (a) 70; 72   (b) 95; 99
   (c) 78; 81   (d) 97; 102

3. (a) 57   (b) 66
   (c) 87   (d) 60
   (e) 92   (f) 103

Exercise 6, p. 16

1. 101   102   105
   101   104   105

2. (a) 136
   (b) 152
   (c) 144
   (d) 163

Exercise 7, p.17

1. (a) 282
   (b) 344
   (c) 298
   (d) 304
   (e) 655
   (f) 333
   (g) 507
   (h) 497

Chapter 3 – Methods for Mental Subtraction

Objectives

♦ Subtract ones or tens from a 2-digit number.
♦ Subtract ones, tens, or hundreds from a 2-digit number.
♦ Subtract two 2-digit numbers.
♦ Subtract a number close to 100 from hundreds.
♦ Subtract a number close to 100 from a 3-digit number.

Notes

In this chapter your students will review mental math strategies for subtraction learned in earlier levels of *Primary Mathematics*, and will learn some new strategies. These strategies include the following:

- Subtract 1, 2, or 3 by counting back.

$$61 - 2 = 59 \qquad \text{Count back: } 60, 59$$
$$302 - 3 = 299 \qquad \text{Count back: } 301, 300, 299$$

- Subtract ones from tens by recalling number bonds for tens.

$$10 - 7 = 3 \qquad\qquad 80 - 7 = 73 \qquad\qquad 380 - 7 = 373$$
$$\overset{\wedge}{70 \quad 10} \qquad\qquad\qquad \overset{\wedge}{370 \quad 10}$$

- Subtract a 1-digit number when there are enough ones by simply subtracting the ones.

$$67 - 2 = 65 \qquad\qquad 867 - 2 = 865$$
$$\overset{\wedge}{60 \quad 7} \qquad\qquad\qquad \overset{\wedge}{860 \quad 7}$$

- Subtract a 1-digit number from a 2-digit number when there are not enough ones

  ⇒ by subtracting from a 10,

$$85 - 7 = 5 + 73 = 78 \qquad\qquad 385 - 7 = 300 + 5 + 73 = 378$$
$$\overset{\wedge}{5 \quad 80} \qquad\qquad\qquad\qquad \overset{\wedge}{305 \quad 80}$$

  ⇒ or by using basic subtraction facts.

$$85 - 7 = 70 + 8 = 78 \qquad\qquad 385 - 7 = 370 + 8 = 378$$
$$\overset{\wedge}{70 \quad 15} \qquad\qquad\qquad\qquad \overset{\wedge}{370 \quad 15}$$

- Subtract tens or hundreds by subtracting tens, using the same strategies used for subtracting ones.

$$419 - 20 = 399 \qquad \text{Count back: } 409, 399$$
$$150 - 70 = 80 \qquad 15 \text{ ones} - 7 \text{ ones} = 8 \text{ ones} \rightarrow 15 \text{ tens} - 7 \text{ tens} = 8 \text{ tens}$$
$$850 - 70 = 780 \qquad 85 \text{ ones} - 7 \text{ ones} = 78 \text{ ones} \rightarrow 85 \text{ tens} - 7 \text{ tens} = 78 \text{ tens}$$

- Subtract a 2-digit number

  ⇒ by subtracting the tens and then the ones, using the strategies already learned.

  $$85 - 47 = 85 - 40 - 7 = 45 - 7 = 38$$

- Subtract a number close to 100 by subtracting 100 and then adding back the difference (this strategy is new here).

  $$485 - 98 = 485 - 100 + 2 = 385 + 2 = 387$$

For students who have not done *Primary Mathematics* before, to start at this level and learn all these mental math strategies for the first time may be a bit overwhelming. If so, don't let your student get hung up on this chapter. Introduce a few strategies, work on those, and move on, coming back to this the lessons in this unit, perhaps once a week, and then spend time reviewing the lesson throughout the week as you go on to other topics. Your student can always add and subtract using the standard algorithm until he is more comfortable with mental math.

Material

- Place-value discs
- 10 x 10 grids (Appendix p. a17)
- Hundred-chart
- Counters or coins for each player
- Four sets of number cards from 0-9 and four extra 0 cards
- 4 number cubes, one labeled with 100, 300, 600, 600, 800, 900, one with 0, 10, 20, 30, 40, 50, one with 5-9, and one with −97, +97, −98, +98, −99 and +99.
- Mental Math 11-20 (Appendix)

(1) Review: Subtract ones or tens from a 2-digit number

Activity

Have your student solve the following problems mentally. Discuss possible strategies. Use number bonds and, if needed, place-value discs to illustrate the possible strategies.

⇒ 12 − 7

Your student may give the answer from having memorized the subtraction fact. Remind him that when there are not enough ones, we can subtract from the ten, and then add in the ones. 10 − 7 + 2 = 5.

Or, since 7 is 2 and 5, subtract 2 from 12, and then 5 more from 10. Essentially, we are subtracting the difference between 7 and 2 from 10.

Or, subtract 10, and then add back in 3 (since by subtracting 10 instead of 7, we subtract 3 too many).

⇒ 30 − 7

Take 7 away from one of the tens. Give your student plenty of practice subtracting from tens.

⇒ 32 − 7

Subtract 7 from the 30, and then include the 2 ones. Or subtract 10 from 32 and add back in 3.

Or, subtract 2 from 32, and then 5 more. Essentially, we are subtracting the difference between 7 and 2 from the tens.

⇒ 32 − 9

9 is one less than 10. If we subtract 10, we subtract one too many. Subtract ten and add back in 1.

⇒ 70 − 50

Subtract tens the same way as subtracting ones, but the answer is tens, rather than ones.

⇒ 72 − 50

First subtract the tens, and then include the 2 ones.

Practice

Task 1, p. 18

Workbook

Exercise 8, pp. 18-19 (Answers p. 24)

Reinforcement

Mental Math 11-12

$12 − 7 = 3 + 2 = 5$

12
/ \
2 10

$12 − 7 = 12 − 2 − 5$
$= 10 − 5$
$= 5$

$12 − 7 = 12 − 10 + 3$
$= 2 + 3$
$= 5$

$30 − 7 = 23$
/ \
20 10

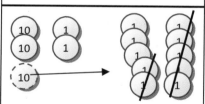

$32 − 7 = 23 + 2 = 25$
/ \
2 30

$32 − 7 = 32 − 2 − 5 = 25$

$32 − 9 = 32 − 10 + 1 = 23$

$70 − 50 = 20$

$72 − 50 = 20 + 2 = 22$
/ \
2 70

---

1. (a) 36   (b) 18   (c) 61
   (d) 20   (e) 50   (f) 40
   (g) 21   (h) 58   (i) 45

(2) Review: Subtract ones, tens, or hundreds

Discussion

Concept page 17

Point out that we subtract tens from tens, ones from ones, and hundreds from hundreds. Illustrate the problems on this page with actual place-value discs if necessary.

Activity

Have your student solve the following problems mentally. Discuss possible strategies. Use place-value discs and/or number bonds to help your student understand the strategies as needed.

⇒ 240 – 6

Subtract 6 from one of the tens. We are left with one less ten and 4 ones. Give your student plenty of practice in subtracting ones from tens.

$$240 - 6 = 230 + 4 = 234$$
$$230 \quad 10$$

⇒ 243 – 6

Again, subtract 6 from one of the tens to get 4. But this time we need to also include the 3 ones. So the answer has one less ten, and the ones is 4 + 3.

Or subtract 3, and then another 3.

$$243 - 6 = 240 - 6 + 3$$
$$= 234 + 3 = 237$$
$$3 \quad 240$$

$$243 - 6 = 243 - 3 - 3$$
$$= 240 - 3$$
$$= 237$$

⇒ 400 – 7

Subtract 7 from a ten. Count down 1 ten from 400: 390. Then add in the answer to 10 – 7.

$$400 - 7 = 393$$
$$390 \quad 10$$

⇒ 403 – 7

Subtract 7 from 400 and then add the 3 ones. Or subtract 3, and then 4 more, since 7 is 3 and 4.

$$403 - 7 = 393 + 3 = 396$$
$$3 \quad 400$$

⇒ 710 – 30

Since 71 ones – 3 ones is 68, then 71 tens – 3 tens is 68 tens. Or, we can subtract 30 from 700, and then add 10.

$$710 - 30 = 680$$
$$10 \quad 700$$

⇒ 812 – 300

Subtract hundreds from hundreds.

$$812 - 300 = 512$$
$$12 \quad 800$$

Practice

Task 2, p. 18

Workbook

Exercise 9, pp. 20-21 (Answers p. 24)

Reinforcement

Mental Math 13-14

2. (a) 223   (b) 197   (c) 403
   (d) 720   (e) 380   (f) 460
   (g) 300   (h) 300   (i) 600
   (j) 342   (k) 353   (l) 608

(3) Review: Subtract two-digit numbers mentally

Activity

Have your student solve the following problems mentally. Discuss possible strategies. Use place-value discs and/or number bonds to illustrate the possible strategies.

⇒ $67 - 35$

From the place-value discs, we can easily see that we can subtract these numbers by simply subtracting tens from tens and ones from ones.

$$67 - 35 = 32$$

⇒ $62 - 35$

We can still subtract tens from tens, but there are not enough ones, so we then need to subtract ones from a ten.

We can first subtract the tens, to get another two digit number, and then the ones, using the same strategies we used to subtract a 1-digit number to a 2-digit number.

Rather than writing down an intermediate number, we can find each digit from left to right:

1. Find the tens. Look ahead to the ones — will subtracting the ones decrease the tens? Yes, so subtract the tens and subtract 1 more ten. The tens digit will be 2.
2. Find the ones: $10 - 5 + 2 = 7$ or $12 - 2 - 3 = 7$.

$$62 \xrightarrow{-30} 32 \xrightarrow{-5} 27$$

$$62 - 35 = \mathbf{2}\_$$
$$\downarrow$$
$$62 - 35 = \mathbf{27}$$

Practice

Tasks 3-4, p. 18

Workbook

Exercise 10, pp. 22-23 (Answers p. 24)

3. 31

4. (a) 41   (b) 71   (c) 30
   (d) 7   (e) 17   (f) 19

Reinforcement

Mental Math 15

Game

Material: A hundred-chart, counters or coins for each player, four sets of number cards from 0-9, and four extra 0 cards.

Procedure: Shuffle the cards and set them face down. Players take turns drawing 4 cards and arranging them into two numbers less than 100. If a 0 is drawn, it can be used as a ten to make a 1-digit number; otherwise both numbers are 2-digit numbers. For example, 3, 6, 4, and 0 are drawn. The numbers could be 40 and 36, 60 and 34, 63 and 4, etc. The player subtracts the smaller number from the larger and covers up the answer on his hundred chart with a counter or coin. The first player to get 3 in a row on the chart wins.

(4) Subtract a number close to 100 from hundreds

Activity

Write the following problems and discuss strategies for solving them mentally, using number bonds to illustrate the process, and 10x10 grids with extra unit squares.

⇒ 100 − 98

$$100 - 98 = 2$$

100 is just 2 more than 98.

⇒ 200 − 98

Subtract from 100 here as well. There is 1 less hundred, and 2 more.

Show two 10 x 10 grids. Ask your student what would happen if we took away 100 instead of 98. We would be taking away 2 too many. So we can subtract 100, and then add back in 2.

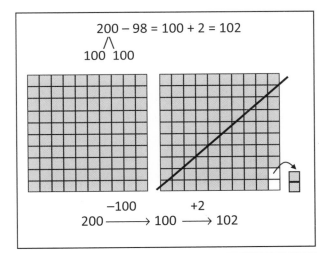

⇒ 900 − 99

Count back 100 and count on 1.

$$900 - 99 = 800 + 1$$
$$= 801$$

⇒ 600 − 97

Ask your student what we need to add back in if we subtract 100. We can count back 100 and then count on 3.

$$600 - 97 = 500 + 3$$
$$= 503$$

⇒ 300 − 92

Count back 100 and add 8.

$$300 - 92 = 200 + 8$$
$$= 208$$

Practice

Tasks 5-6, p. 19

Workbook

Exercise 11, p. 24 (Answers p. 24)

Reinforcement

Mental Math 16

5. 201

6. (a) 101   (b) 301   (c) 801
   (d) 602   (e) 402   (f) 702

(5) Subtract a number close to 100 from a 3-digit number

Activity

Write the following problems and discuss strategies for solving them mentally, using number bonds to illustrate the process.

$\Rightarrow$ 603 – 98

Split 603 into 503 and 100 and subtract 98 from 100. This is essentially the same as subtracting 100 and adding 2.

$$603 - 98 = 503 + 2 = 505$$
$$503 \quad 100$$

$$603 \xrightarrow{-100} 503 \xrightarrow{+2} 505$$

$\Rightarrow$ 683 – 99

The easiest way to solve this is to count back 1 hundred and then count on 1.

$$683 - 99 = 683 - 100 + 1$$
$$= 583 + 1$$
$$= 584$$

$\Rightarrow$ 422 – 96

Ask your student what we would have to add back in if we subtracted 100 from 422 to get the correct answer. We can count back 100 and then add 4.

$$422 - 96 = 422 - 100 + 4$$
$$= 322 + 4$$
$$= 326$$

Practice

Tasks 7-9, p. 19

Workbook

Exercise 12, p. 25 (Answers p. 24)

Reinforcement

Mental Math 17

7. 105

8. (a) 3   (b) 209   (c) 506
   (d) 206   (e) 303   (f) 608

9. (a) 141   (b) 247   (c) 613
   (d) 223   (e) 456   (f) 832

Activity

Material: 4 number cubes, one labeled with 100, 300, 600, 600, 800, 900, one with 0, 10, 20, 30, 40, 50, one with 4, 5, 6, 7, 8, 9, and one with –97, +97, –98, +98, –99 and +99.

Procedure: Throw the cubes, form a 3-digit number from three of them, and then perform the operation shown on the fourth cube.

If your student likes to compete for speed, throw the four cubes and see who, you or your student or other players, can find the answer mentally first.

(6) Practice

Practice

Practice B, p. 20

Use this practice to review mental math strategies and to provide practice with word problems.

For the word problems, if your student has difficulty, ask her to determine whether two parts are given and the whole must be found, or whether the whole and one part are given and a missing part must be found. If the problem involves comparing two numbers, ask her which is larger. Guide her in drawing a number bond or simple drawing with the information so that she can determine what equation to use.

Reinforcement

Mental Math 18-19

Extra Practice, Unit 7, Exercise 3, pp. 107-110

Game

Material: Hundred-chart and markers of a different color for each person.

Procedure: Each player takes turns. The first player circles a number. The next player circles another number. For each succeeding turn, the player must circle a number that is the difference between any two numbers already circled. Play continues until no more numbers can be circled.

Test

Tests, Unit 7, Chapter 3, A and B, pp. 9-12

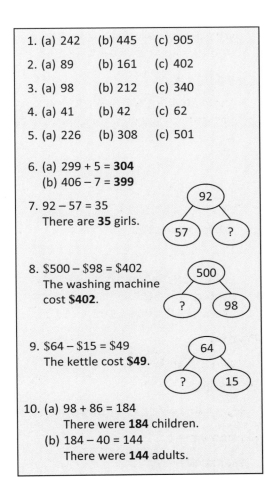

1. (a) 242   (b) 445   (c) 905

2. (a) 89   (b) 161   (c) 402

3. (a) 98   (b) 212   (c) 340

4. (a) 41   (b) 42   (c) 62

5. (a) 226   (b) 308   (c) 501

6. (a) 299 + 5 = **304**
   (b) 406 − 7 = **399**

7. 92 − 57 = 35
   There are **35** girls.

8. $500 − $98 = $402
   The washing machine cost **$402**.

9. $64 − $15 = $49
   The kettle cost **$49**.

10. (a) 98 + 86 = 184
    There were **184** children.
    (b) 184 − 40 = 144
    There were **144** adults.

Workbook

Exercise 8, pp.18-19

1. (a) 15   (b) 10   (c) 71
   (d) 13   (e) 50   (f) 32

2. (a) 21   (b) 64
   (c) 31   (d) 51
   (e) 72   (f) 91
   (g) 18   (h) 58
   (i) 78   (j) 46
   (k) 39   (l) 66
   (m) 46   (n) 28
   (o) 95   (p) 78

3. (a) 18   (b) 56
   (c) 65   (d) 43
   (e) 22   (f) 34
   (g) 71   (h) 87

4. (a) 10   (b) 20
   (c) 30   (d) 40
   (e) 10   (f) 10
   (g) 0   (h) 10

5. (a) 21   (b) 33
   (c) 27   (d) 18
   (e) 14   (f) 25
   (g) 59   (h) 12

Exercise 9, pp. 20-21

1. (a) 872   (b) 934
   (c) 412   (d) 262
   (e) 103   (f) 653

2. (a) 442   (b) 678
   (c) 888   (d) 556
   (e) 228   (f) 944
   (g) 718   (h) 137

3. (a) 503   (b) 757
   (c) 121   (d) 327
   (e) 230   (f) 806
   (g) 632   (h) 428

4. (a) 469   (b) 658
   (c) 186   (d) 283
   (e) 377   (f) 545
   (g) 198   (h) 771

5. (a) 100   (b) 200
   (c) 700   (d) 200
   (e) 400   (f) 300
   (g) 100   (h) 300

6. (a) 433   (b) 89
   (c) 153   (d) 394
   (e) 235   (f) 227
   (g) 286   (h) 168

Exercise 10, pp. 22-23

1. (a) 2   (b) 80   (c) 52

2. (a) 48; 43   (b) 25; 21
   (c) 11; 8   (d) 23; 17

3. (a) 42   (b) 12
   (c) 23   (d) 9
   (e) 18   (f) 18

4. (clockwise from top) 89; 55; 58; 98

5. (a) 23   (b) 89
   (c) 74   (d) 37
   (e) 17   (f) 92

Exercise 11, p. 24

1. (a) 201
   (b) 401
   (c) 601
   (d) 701
   (e) 302
   (f) 502
   (g) 202
   (h) 802

Exercise 12, p. 25

1. (a) 81
   (b) 203
   (c) 457
   (d) 749
   (e) 107
   (f) 369
   (g) 682
   (h) 534

Review 7

Review

Review 7, pp. 21-23

This is a cumulative review and includes material from earlier levels of *Primary Mathematics*.

Your student should be able to determine whether to use mental math strategies or to use the addition or subtraction algorithm. Specific mental math strategies have not been taught for all of these problems. You should allow her to use the standard algorithms for addition and subtraction when needed.

For example, mentally adding a 2-digit number to a 3-digit number has not specifically been taught, but your student may want to add 333 + 78 in 2(b) mentally by first adding tens and then ones. Problem 15 involves subtraction from 1000. This was not addressed in the textbook, but if you did the enrichment activity on p. 5 of this guide, he may want to do this one mentally. He could simply count up from 58 by ones and tens. Problem 18 involves adding two 3-digit numbers; he may prefer to add them using the standard algorithm, or he may extend some of the strategies learned in this chapter and add them mentally.

Problem 17 involves division by 3, which was taught in *Primary Mathematics* 2A. You can save this problem until after the next unit, if your student needs to review multiplication and division concepts.

Workbook

Review 8, pp. 26-29

(This is Review 8, instead of 7, even though it is unit 7, because there was an extra review at the end of the *Primary Mathematics* 2A workbook.)

Enrichment

Mental Math 20

Test

Tests, Units 1-7, Cumulative, A and B, pp. 13-16

1. (a) 386   (b) 327   (c) 210
2. (a) 250   (b) 411   (c) 500
3. (a) 507   (b) 230   (c) 498
4. (a) 731   (b) 455   (c) 178
5. (a) 202   (b) 359   (c) 502
6. (a) pencil
   (b) 1 cm
7. 12
8. 9   8
9. 495   594
10. (a) 745
    (b) seven hundred forty-five
11. 255 + 45 = 300
    She had **300** eggs at first.
12. 185 + 28 = 213
    He made **213** cheese tortillas.
13. 135 cm + 29 cm = 164 cm
    David is **164 cm** tall.
14. 65 lb − 19 lb = 46 lb
    His sister weighs **46 lb**.
15. 1000 − 958 = 42
    42 people were not present.
16. 105 + 95 = 200
    She sold **200** eggs on Monday.
17. 18 ft ÷ 3 = 6 ft
    Each piece is **6 ft** long.
18. (a) 215 + 285 = 500
        He sold **500** copies both days.
    (b) 285 − 215 = 70
        He sold **70** more on Sunday.
19. (a) 86 kg + 54 kg = 140 kg
        he sold **140 kg** altogether.
    (b) 200 kg − 140 kg = 60 kg
        he had **60 kg** left.
20. (a) $488 - $85 = $403
        The television cost **$403**.
    (b) $403 + $488 = $891
        The total cost is **$891**.

Workbook

Review 8, pp. 26-29

1. 769
   405
   30
   60

2. (a) 350
   (b) 704

3. 90   56   500   680
   55   93   32   100

4.

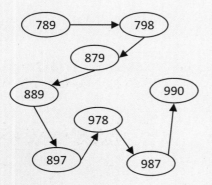

5. 99 + 26 = 125
   Matthew has **125** seashells now.

6. $300 − $98 = $202
   She spent **$202**.

7. 135 + 45 = 180
   Emily collected **180** stickers.

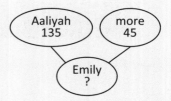

8. $200 − $85 = $115
   She saved **$115** in January.

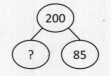

9. 650 g − 200 g = 450 g
   The pear weighs **450 g**.

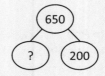

10. 350 m + 250 m = 600 m
    The post office was **600 m** from his house.

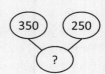

Unit 8 – Multiplication and Division

Chapter 1 – Multiplying and Dividing by 4

Objectives

- Count by 4's.
- Relate the associated facts 4 x _____ and _____ x 4.
- Build the multiplication table for 4.
- Use related facts to find unknown facts.
- Relate division by 4 to multiplication by 4.
- Solve word problems involving multiplication or division by 2, 3, or 4.

Notes

In *Primary Mathematics* 2A students learned the multiplication and division facts for 2 and 3. They should be familiar with these facts. Incorporate review of them with the new facts that will be learned in this unit.

Both multiplication and division are associated with the part-whole concept. Given the number of equal parts and the number in each part (its value), we can multiply to find the whole (total). For any given multiplication situation we can write two related equations.

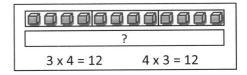

$3 \times 4 = 12$ $4 \times 3 = 12$

In *Primary Mathematics* we study two division situations:

Sharing:
A total amount (the whole) is shared into a given number of groups (parts). Divide the total by the number of parts to find the number (value) in each part.

Grouping:
A total amount (the whole) is grouped into equal groups (parts). The number that goes into each part (value of each part) is given. Divide the total by the number that goes into each part to find the number of parts.

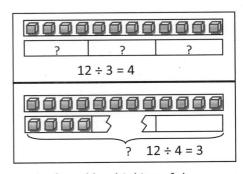

$12 \div 3 = 4$

$12 \div 4 = 3$

Multiplication and division are related. The answer to $12 \div 4$ can be found by thinking of the number times 4 that equals 12.

In this chapter your student will start by building the multiplication tables of 4 based on the idea of "4 more than." She will use the multiplication facts for 4 to learn the division facts.

Material

- 1-40 number board (Appendix p. a18)
- Counters
- Multilink cubes
- Hundred-chart
- Number cubes
- 4 sets of number cards 1-10
- Number cube
- Appendix pages a18-a21
- Mental Math 21-25 (Appendix)

(1) Count by 4's to multiply 4

Discussion

Concept page 24

Ask your student how many stickers there are in each row. The number at the end of each row shows the total stickers, counting from the top row to the bottom. They show what we get if we count by fours. Cover up the numbers with an index card and have your student count by 4's as you uncover the numbers by sliding the card down.

Activity

Use the 1-40 number board in the appendix or a hundred-chart. Ask your student to cover up the numbers he would land on if counting by 4's (multiples of 4). Then have him tell you the covered numbers. Then have him practice counting by fours both forwards and backwards without the chart.

1	2	3	◯	5	6	7	◯	9	10
11	◯	13	14	15	◯	17	18	19	◯
21	22	23	◯	25	26	27	◯	29	30
31	◯	33	34	35	◯	37	38	39	◯

Discussion

Concept page 25

12; 12
28; 28

Remind your student that we use a multiplication sign when we have equal groups. For the first 3 rows of stickers, we have a group of 4, multiplied 3 times. Ask her how many stickers are in the 3 rows. We can show what we want to find with the expression **4 x 3**. To find the answer, we can count by 4's.

Practice

Task 1, p. 25

1. (a) 16
   (b) 36

Workbook

Exercise 1, pp 30-32 (Answers p. 35)

Reinforcement

Provide your student with multilink cubes in 10 groups of 4. Tell him to set out 6 groups of 4. Write an addition expression and have him find the answer. Then write the multiplication expression and have him again tell you the answer.

Write an expression with 4 x __, such as 4 x 8. Get your student to show the correct number of fours with the cubes and then tell you the answer. Repeat as needed.

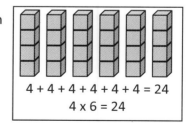

4 + 4 + 4 + 4 + 4 + 4 = 24
4 x 6 = 24

To count by fours, let your student hold up a finger for each four, first on one hand, then the next, to keep track of how many fours he has counted. Until she has the facts memorized, she can count by fours until she gets to the correct number of fingers. She can also count on by fours from a fact she knows. For example, if she knows that 4 x 5 = 20, she can hold up 5 fingers, say 20, and then count by fours for the sixth and seventh finger. Eventually she will not need to use fingers.

(2) Build multiplication tables for 4

Activity

Create two 4 by 3 rectangular arrays with the multilink cubes. Separate one into 3 columns of 4 and write an addition equation, and then the multiplication equation 4 x 3 = 12, asking your student for the answer. 4 multiplied by 3 is 12. Then, separate the other into rows. Write the addition equation, and then 3 x 4 = 12. 3 multiplied by 4 is also 12.

Go back to the 3 columns. Tell your student that we could have written 3 x 4 = 12 for this, thinking of the numbers as 3 groups of 4. "4 multiplied by 3 is 12" is the same as "3 groups of 4 is 12". There is no rule that says we have to write the multiplication expression for this as either 4 x 3 or 3 x 4.

Then, have your student look at the 4 rows of 3 again. This is "3 multiplied by 4" but it is also "4 groups of 3". We can write this as 3 x 4 or 4 x 3.

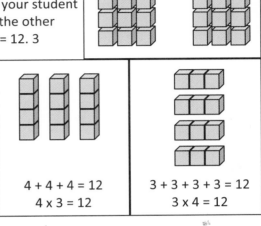

4 + 4 + 4 = 12
4 x 3 = 12

3 + 3 + 3 + 3 = 12
3 x 4 = 12

The two situations are not the same. One has 3 groups of 4, and the other 4 groups of 3. But since the answer is the same either way, we could solve 4 multiplied by 3, 4 x 3, by either counting by 3 four times, or by counting by 4 three times.

Show 3 columns of 4. Add another column of 4. Write the expressions 4 x 3 = 12, then 4 x 4 = ?. Tell your student that if we know that 4 x 3 = 12, then we can find 4 x 4 by simply adding 4 more to the answer of 4 x 3. Similarly, if we know that 4 x 4 = 16, then we can find 4 x 3 by subtracting 4 from the answer to 4 x 4.

Ask, "If 4 x 5 is 20, what is 4 x 6?" Repeat with some other examples. You can even include larger numbers, for example, "If 4 x 50 is 200, what is 4 x 51?" Include some where your student must subtract 4 from the given answer, such as, "If 4 x 10 is 40, what is 4 x 9?".

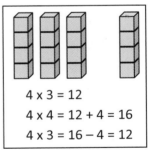

4 x 3 = 12
4 x 4 = 12 + 4 = 16
4 x 3 = 16 − 4 = 12

Practice

Tasks 2-5, pp 26-27

For task 4 your student should write the answers. She can find each answer by adding 4 to the previous answer. You can have her write the equations out, or make a copy of table on p. a19 in the appendix.

Workbook

Exercise 2, pp. 33-35 (Answers p. 35)

2. 32		
3.	28	36
	20	28	36
4.	8	
	12	12
	16	16
	20	20
	24	24
	28	28
	32	32
	36	36
	40	40
5. 24		

(3) Memorize multiplication facts for 4

Activity

Set out some cubes as shown here. Write the total number of cubes under each column and write the two equations **2** x 3 = 6 and **4** x 3 = 12. Ask your student if he notices anything. To find 2 x 3 we can double 3. If we double that answer, then we get the answer to 4 x 3. So we can find 4 x 3 by doubling 3 twice. Now ask him to double 6 (6 + 6 = 12) and double the answer (12 + 12 = 24). Let him look at the multiplication table he made for 4 and see if the answer to 4 x 6 (or 6 x 4) is 24. Give him some other numbers between 1 and 10 to double twice.

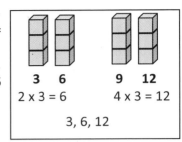

3   6   9   12

2 x 3 = 6   4 x 3 = 12

3, 6, 12

Choose one or more of the following activities, whichever works best for your student.

⇒ Show your student how to fill the first multiplication chart on appendix p. a20 and have her do the other two independently (cut them apart so that she is not using one chart to fill out the other).

⇒ Have your student say or "chant" the multiplication facts periodically during the next few lessons - "one four is four, two fours is eight, three fours is twelve."

⇒ Make a set of fact cards for the multiplication facts of 4. Shuffle and show your student each card. If he gives the correct answer, put it in his pile; if not, tell him the correct answer and put it in your pile. Repeat with your pile until he has all the cards.

⇒ Use four sets of number cards 1-10. Shuffle and place face down. Your student turns over the cards one at a time and supplies the fact for 4 times the number drawn. If correct, he places the cards in one pile; if wrong, he places them in a second pile. Repeat with the second pile until he has all the cards.

⇒ Make a game board with multiples of 4 in random order. You can do this on the back of a laminated hundred-board. Make two sets of discs with 1-9 written on them, such as writing the numbers on counting discs. Turn them all upside down and mix them up. Your student turns over one disc at a time and lays the disc on the number multiplied by 4. For example, she turns

4	32	8	12	4
28	20	8	24	12
16	28	32	40	20
40	24	36	16	36

over 9. She puts it on one of the 36 squares. You can time her and let her see how fast she puts the discs on the correct number and if she can beat her previous time. She can also see how soon she gets 3 or 4 in a row.

⇒ Game
Material: 4 sets of number cards 1-10, number cube marked with 2, 3, 4, 4, and 4.
Procedure: Shuffle the cards and place the deck face-down in the middle. Each player draws a card, throws the number cube, and multiplies the number on the card by the number on the cube. The player with the greatest answer gets all the cards. If both players have the same answer then the one with the greatest factor gets the cards. The player with the most cards at the end wins.

Reinforcement

Mental Math 21-22

(4) Solve word problems

Activity

Discuss some word problems that involve multiplication by 4 with your student, such as the following. Write them down and ask him to read them out loud. Ask him what he is supposed to find. Get him to tell you whether there are equal groups, how many equal groups, and how much is in each group. Have him write an equation for each problem. Allow him to act out the problem with connect-a-cubes or other objects he can put into groups or he can make drawings if necessary. Sample drawings are shown for several problems below. Get him to tell you the answer to the problem in a complete sentence that includes what the problem asked him to find.

⇒ Jeremy bought 4 books. Each book cost $6. How much did he pay for the books? (4 x $6 = $24, He paid $24 for the books.)

⇒ A plastic bag can hold 4 apples. How many apples can 7 plastic bags hold? (4 x 7 = 28; 7 bags can hold 28 apples.)

⇒ Megan wants to wrap 6 presents. Each present needs 4 feet of ribbon. How many total feet of ribbon does she need to wrap all 6 presents? (6 x 4 ft = 24 ft; She needs 24 ft of ribbon.)

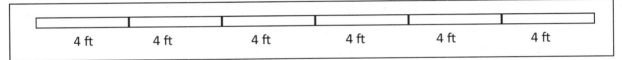

⇒ There are 35 children at a party. There are 4 packages of large balloons with 8 balloons in each package. Are there enough balloons for each child? (8 x 4 = 32; There are not enough balloons.)

⇒ There are 4 monkeys in a zoo. Each monkey eats 3 bananas in the morning and 4 bananas in the afternoon.

How many bananas do they eat in the morning? (4 x 3 = 12; They eat 12 bananas in the morning)

How many bananas do they eat in the afternoon? (4 x 4 = 16; They eat 16 bananas in the afternoon.)

How many bananas do they eat altogether? (12 + 16 = 28 or 4 x 7 = 28; They eat 28 bananas altogether.)

Continue to help your student learn the multiplication facts for 4. Include a review of the multiplication facts for 2 and 3.

Workbook

Exercise 3, pp. 36-39 (Answers p. 35)

(5) Relate division by 4 to multiplication by 4

Activity

Set out a pile of 24 counters. Also set out 4 small bowls or plates or draw 4 circles. Make up a story, such as, "There are 24 cookies. We want to divide them up among 4 children so that each child gets the same number of cookies. How many cookies will each child get?" Write the division problem 24 ÷ 4 = ___. Draw the number bond as shown at the right. Without letting your student divide up the counters, ask her how many counters will she have to put in each bowl. If she gives the correct answer, ask her how she found the answer. We will end up with equal groups. So if we can remember from the multiplication facts what number times 4 is 24, then we know how many go into each group. Have her check her answers by dividing up the counters.

You can draw an arrow diagram to show the relationship between division and multiplication.

Put the counters back into a pile. Tell your student that earlier we put them into equal groups and found how many were in each group. Remind him that 24 ÷ 4 could also mean taking 24 things and grouping them by 4 to find out how many groups there are. Make up another story, such as "There are 24 cookies and some children. Each child got 4 cookies. How many children were there?" Let him group the counters by 4 to show that there are 6 children. Write the division problem. Tell him that instead of finding how many are in 4 groups, we found how many 4's are in 24. The answer is the same as in the previous problem. We can still think of what number x 4 = 24 to find the answer.

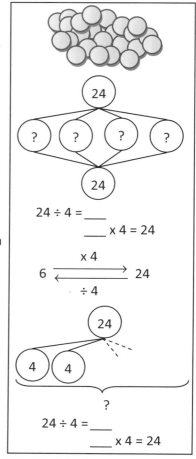

Put the counters into a 4 by 6 array. Have your student write two multiplication problems to represent "4 rows x 6 in each row is 24 total" and "6 columns x 4 in each column is 24 total." Then guide her in writing two division problems to represent the total divided by 4 rows is 6 in each row and the total divided by 6 columns is 4 in each column. Point out that 24 ÷ 4 = 6 could also mean 24 grouped by 4's is 6 groups, or there are 6 fours in 24. 24 ÷ 6 = 4 could also mean 24 grouped by 6's is 4 groups, or there are 4 sixes in 24.

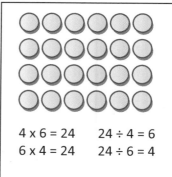

4 x 6 = 24	24 ÷ 4 = 6
6 x 4 = 24	24 ÷ 6 = 4

Practice

Tasks 6-8, pp. 27-28

Reinforcement

Have your student fill in the blanks for the problems on appendix p. a21.

6. (a) 3, 3	(b) 3, 3, 3	
7. 2	8	
5, 5	7, 7	
8. (a) 1	(b) 4	(c) 9
(d) 6	(e) 3	(f) 10

(6) Memorize division facts for 4

Activity

By now, your student should begin to recognize whether a given number is a multiple of 2, 3, or 4 up to ten times.

Use number cards with the numbers 2, 3, 4, 6, 8, 9, 10, 12, 14, 15, 16, 18, 20, 21, 22, 24, 27, 28, 30, 32, 36, and 40. Shuffle. Draw one at a time and ask your student to supply as many multiplication facts she can for each.

Randomly say a number between 1 and 40 and ask if it can be divided evenly by 4.

Help your student memorize the division facts for 4, using fact cards, games, or other methods, such as adaptations of the activities given for multiplication. Include a review of the division facts for 2 and 3. You can also try one of the following games.

Game

Material: A game board with the numbers 1 through 10. (You can make the game board as large as you want by repeating the numbers 1-10. You can use the back of a laminated hundred chart. Write the numbers 1-10 randomly in the squares using dry-erase or wet-erase markers. Each time you play the game change the placement of the numbers.) A set of division cards for the division facts for 2, 3, and 4. (Use index cards and write the division expressions on them: 40 ÷ 4, 36 ÷ 4, ... , 30 ÷ 3, etc.) Counters of different colors for each player.

Procedure: Shuffle the cards and place them face down. Players take turns drawing a card and placing a marker on the board for the answer. The winner is the first to get three markers in a row.

Game

Material: A 5 x 5 game board with the numbers 2, 3, 4, 6, 8, 9, 10, 12, 12, 14, 15, 16, 16, 18, 20, 21, 22, 24, 24, 27, 28, 30, 32, 36, and 40 randomly placed. (You can use the back of a hundred board and write the numbers in the squares. For a larger game board, use all 100 squares and repeat the numbers. Each time you play the game, change the placement of the numbers.) A number cube with the 2, 2, 3, 3, 4 and 4. Counters of different colors for each player, or coins.

Procedure: The players take turns throwing the number cube and placing a marker on a number that would be 1 times, 2 times, 3 times, or 4 times the number thrown. The winner is the first to get three markers in a row. Your student may notice that 28 can count as a multiple of 2, or 36 as a multiple of 3.

Reinforcement

Mental Math 23-24

(7) Solve word problems and practice

Discussion

Task 9, p. 28

Write the problems on a white board so that your student does not see the solution before contributing to the discussion. Ask her to read the problem out loud and then tell you what she must find. Let her act out these problems if needed, or draw a diagram. 9(a) is a sharing problem and she could draw a big circle and write $32 in it, and then 4 smaller circles for each day. 9(b) is a grouping problem, which is harder to diagram, but if she starts the diagram she can see again that she needs to divide. You can say that she needs to find how many $4's there are in $40.

Point out that in both problems we are given a total. In the first we are given equal groups (days), and have to find the amount of money for each day. In the second we are given how much goes in each group and need to find the number of groups (days). When we are given a total, we will need to subtract or divide. If we have unequal parts, we need to subtract. If we have equal parts or are making equal parts, we need to divide.

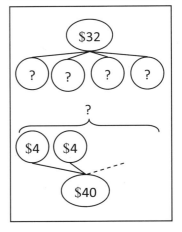

9. (a) $8; $8
   (b) 10; 10

Practice

Practice A, p. 29

Encourage your student to write the measurement units with the answers, not just the numerical answer (e.g., $, m, kg, in. ...). Also encourage him to draw diagrams to help him determine what operation (x or ÷) is needed.

Problem 11: Your student cannot solve this from having memorized division facts for 4. If she understands well the relationship between multiplication and division, she might realize she can think of what number times 9 is 36. Otherwise, she may have to use counters to solve this problem.

Workbook

Exercise 4, pp. 40-42 (Answers p. 35)

Reinforcement

Extra Practice, Unit 8, Exercise 1, pp. 115-118

Mental Math 25

Test

Tests, Unit 8, Chapter 1, A and B, pp. 21-24

1. (a) 12   (b) 28   (c) 8
2. (a) 1   (b) 8   (c) 4
3. (a) 24   (b) 40   (c) 32
4. (a) 2   (b) 5   (c) 10
5. (a) 9   (b) 3   (c) 6

6. 1 taxi: 4 passengers
   5 taxis: 4 x 5 = **20** passengers.

7. Group 16 kg by 4's.
   16 kg ÷ 4 = 4 kg
   There are **4 kg** in each bag.

8. 1 child: 4 books
   6 children: 4 x 6 = 24 books
   They borrowed **24** books.

9. 4 T-shirts: $40
   1 T-shirt: $40 ÷ 4 = $10
   Each T-shirt cost **$10**.

10. 1 set: 8 m
    4 sets: 8 x 4 = 32 m
    She used **21 m** of cloth.

11. Group 36 by 9's
    36 ÷ 9 = 4   (Think: ? x 9 = 36)
    There are **4** boxes.

Workbook

Exercise 1, pp. 30-32

1. 4; 8; 12; 16; 20; 24; 28; 32; 36; 40

2. (a) 4
   (b) 8
   (c) 12
   (d) 16
   (e) 20
   (f) 24
   (g) 28
   (h) 32
   (i) 36
   (j) 40

Exercise 2, pp. 33-35

1. (a) 8
      8
   (b) 12
      12
   (c) 28
      28
   (d) 36
      36

2. (a) 12   (b) 24
   (c) 20   (d) 40

3. 16   32
   28   36
   20   28

Exercise 3, pp. 36-39

1. $2 \times 4 \rightarrow 8$ $9 \times 4 \rightarrow 36$
   $3 \times 4 \rightarrow 12$ $7 \times 4 \rightarrow 28$
   $6 \times 4 \rightarrow 24$ $5 \times 4 \rightarrow 20$
   $8 \times 4 \rightarrow 32$ $10 \times 4 \rightarrow 40$

2. $4 \times 5 = 20$
   There were **20** trees altogether.

3. $4 \times 6 \text{ cm} = 24 \text{ cm}$
   The total length is **24 cm**.

4. $4 \times 3 = 12$
   They caught **12** fish altogether.

5. $4 \times 2 = \mathbf{8}$ $14 = 7 \times 2$
   $5 \times 3 = \mathbf{15}$ $21 = 3 \times 7$
   $2 \times 9 = \mathbf{18}$ $20 = 4 \times 5$
   $4 \times 8 = \mathbf{32}$ $18 = 6 \times 3$
   $6 \times 4 = \mathbf{24}$ $10 = 2 \times 5$
   $10 \times 4 = \mathbf{40}$ $9 = 3 \times 3$
   $4 \times 9 = \mathbf{36}$ $24 = 3 \times 8$

6. $\$4 \times 6 = \$24$
   Denisha paid **$24** altogether.

7. $3 \text{ m} \times 9 = 27 \text{ m}$
   She bought **27 m** of cloth.

8. $2 \text{ kg} \times 10 = 20 \text{ kg}$
   She bought **20 kg** of oil.

Exercise 4, pp. 40-42

1. 1   2
   3   4; 4
   5; 5   6; 6
   10; 10   7; 7
   9; 9   8; 8

2. $32 \div 4 \rightarrow 8$
        $24 \div 4 \rightarrow 6$ $20 \div 4 \rightarrow 5$
      $16 \div 4 \rightarrow 4$ $36 \div 4 \rightarrow 9$
        $28 \div 4 \rightarrow 7$ $8 \div 4 \rightarrow 2$
                    $12 \div 4 \rightarrow 3$

3. $36 \div 4 = 9$
   There were **9** children in each row.

4. $\$24 \div 4 = \$6$
   Each kg cost **$6**.

5. $28 \text{ m} \div 4 = 7 \text{ m}$
   Each piece is **7 m** long.

Chapter 2 – Multiplying and Dividing by 5

Objectives

◆ Count by 5's.
◆ Relate the associated facts 5 x _____ and _____ x 5.
◆ Relate division by 5 to multiplication by 5.
◆ Solve word problems involving multiplication or division by 2, 3, 4, or 5.

Notes

In this chapter your student will learn the multiplication and division facts for 5 and solve problems involving multiplication and division by 5 as well as by 2, 3, and 4.

Students already learned to count by 5's in *Primary Mathematics* 1B. Multiplication and division facts for 5 should be easy to learn.

Continue to have your student practice all the multiplication and division facts learned so far.

Material

◆ 1-50 number board (Appendix p. a18)
◆ Counters
◆ Multilink cubes
◆ Hundred-chart
◆ Number cubes
◆ 4 sets of number cards 1-10
◆ Appendix pages a22-a24
◆ Mental Math 26-30 (Appendix)

(1) Count by 5's to multiply by 5

Discussion

Concept page 30

The picture cards are being counted in groups of 5. Ask your student how many cards she could buy with different amounts of dollars up to $10, e.g., "How many cards could you buy for $7?"

Activity

Use the 1-50 number board in the appendix or a hundred chart. Ask your student to circle the numbers he would land on if counting by 5's (multiples of 5). Ask him if he sees any pattern. All the circled numbers end with 5 or 0.

1	2	3	4	5	6	7	8	9	10
11	12	13	14	15	16	17	18	19	20
21	22	23	24	25	26	27	28	29	30
31	32	33	34	35	36	37	38	39	40
41	42	43	44	45	46	47	48	49	50

List all the facts for 5 below the number board. See if he sees any patterns. 5 multiplied by 1, 3, 5, 7, and 9 give answers that end in 5. 5 multiplied by 2, 4, 6, 8, and 10 give answers where the tens digit is half the number 5 is being multiplied by. For numbers which we can divide evenly by 2 with no left-overs (even numbers) we can divide the number by 2 and add a 0 to get the same answer as we get by multiplying by 5. If the number can't be divided evenly by 2 (odd numbers), then 5 times that number is a number that ends in 5.

$$5 \times 1 = 5 \qquad 5 \times 2 = 10$$
$$5 \times 3 = 15 \qquad 5 \times 4 = 20$$
$$5 \times 5 = 25 \qquad 5 \times 6 = 30$$
$$5 \times 7 = 35 \qquad 5 \times 8 = 40$$
$$5 \times 9 = 45 \qquad 5 \times 10 = 50$$

Have your student practice counting by 5's both forward and backward, with and without the chart.

Have your student fill out a copy of the multiplication tables on p. a22 in the appendix.

Discussion

Tasks 1-3, pp. 31-32

Ask your student to write equations for task 3.

Workbook

Exercise 5, p. 43 (Answers p. 41)

Reinforcement

Have your student fill out one or more of the multiplication charts copied from appendix p. a 23 Separate the charts from each other.

1. (a) 15   (b) 40
        15   40

2. 45
     45

3. (a) 5¢ x 6 = 30¢
   (b) $5 x 7 = $35

(2) Learn multiplication facts for 5, solve word problems.

Discussion

Tasks 4-5, p. 32

4. 20	45
	45
5. (a) 35	(b) 30 (c) 5
(d) 25	(e) 10 (f) 50

Remind your student that if a number can be divided evenly by 2, when we multiply it by 5 the answer is the ten of half of the number. For example, the answer to 5 x 6 or 6 x 5 is 30. 3 is half of 6. If the number cannot be divided evenly by 2, the answer is 5 more than the ten half of one less than that number. The answer to 7 x 5 is just 5 more than 30, 3 is half of 7 – 1. So, to find 9 x 5, we can think of 8 x 5, half of 8 is 4, so 8 x 5 is 40, and 9 x 5 is 5 more, or 45. Or we can subtract 5 from 10 x 5.

Activity

Help your student memorize the multiplication facts for 5 using fact cards, games, or other methods, such as adaptations of the activities given in the previous chapter.

Write the following problems and have your student solve them. Ask her to read the problem out loud and give you the answer in a sentence that includes what the problem asked for. She can act out the problem with counters, or draw a diagram, if needed.

⇒ Melanie bought 5 dolls. Each doll cost $6. How much did she pay for the dolls? (5 x $6 = $30; She paid $30 for the dolls.)

⇒ A box can hold 8 cupcakes. How many cupcakes can 5 such boxes hold? (5 x 8 = 40; 5 boxes can hold 40 cupcakes.)

⇒ Cecily wants to make 9 curtains. For each curtain she needs 5 feet of material. How many feet of material does she need? (5 ft x 9 = 45 ft; She needs 45 ft.)

⇒ There are 7 elephants in a zoo. Each elephant eats 5 bundles of hay in the morning and 4 bundles in the afternoon.

  • How many bundles of hay do they eat in the morning? (5 x 7 = 35; They eat 35 bundles of hay in the morning.)

  • How many bundles of hay do they eat in the afternoon? (4 x 7 = 28; They eat 28 bundles of hay in the afternoon.)

  • How many more bundles do they eat in the morning than in the afternoon? (35 – 28 = 7; They eat 7 more bundles of hay in the morning than in the afternoon.)

Workbook

Exercise 6, pp. 44-45 (Answers p. 41)

Reinforcement

Mental Math 26-27

(3) Divide by 5

Activity

Have your student fill out a copy of the chart on appendix p. a24. To divide a number by 5, we can think of what number times 5 gives that number. For example, to divide 35 by 5, think of what number times 5 is 35.

Ask your student to look at the division answers and compare them to the number we are dividing by 5. Does he see any pattern? Ask him to double the number being divided by 5 and compare that to the answer.

To divide a number by 5, we can double it (add the number to itself) and remove the 0.

$35 \div 5 = \underline{}$

$\underline{} \times 5 = 35$

$$7 \xleftarrow[\div 5]{\times 5} 35$$

$5 \div 5 = \quad 1$
$10 \div 5 = \quad 2$
$15 \div 5 = \quad 3$
$20 \div 5 = \quad 4$
$25 \div 5 = \quad 5$
$30 \div 5 = \quad 6$
$35 \div 5 = \quad 7$
$40 \div 5 = \quad 8$
$45 \div 5 = \quad 9$
$50 \div 5 = \quad 10$

35 doubled = 70
$35 \div 5 = 7$

Practice

Tasks 6-7, p. 32

Workbook

Exercise 7, problems 1-2, pp. 46-47 (Answers p. 41)

Reinforcement

Mental Math 28-29

Continue to provide your student with opportunities to learn the math facts for multiplication and division by 2, 3, 4, and 5.

6. 3   8
        8

7. (a) 6   (b) 1   (c) 5
    (d) 2   (e) 10   (f) 9

(4) Solve word problems and practice

Activity

Write the following problems and guide your student in finding the answer. Allow her to act the problem out with counters or draw a diagram, if needed.

⇒ A box of 25 cookies was shared among 5 children equally. How many cookies did each child get? (25 ÷ 5 = 5)

⇒ You have 24 toothpicks.

1. How many equal sized squares can you form with the toothpicks? (6 with sides one toothpick long or 3 with sides two toothpicks long)

2. How many equal sized triangles of equal sides can you form with the toothpicks? (8 with sides 1 toothpick long, 4 with sides 2 toothpicks long)

⇒ A box of 5 toy cars cost $9. Josh bought 20 cars.

1. How many boxes did he buy? (20 ÷ 5 = 4)

2. How much did he pay? (4 x $9 = $36)

Practice

Practice B, p. 33

Workbook

Exercise 7, problems 3-5, p. 48 (Answers p. 41)

Reinforcement

Mental Math 30

Extra Practice, Unit 8, Exercise 2, pp. 119-120

Test

Tests, Unit 8, Chapter 2, A and B, pp. 25-28

Enrichment

⇒ On a test 5 points were given for each correct answer and 3 points subtracted for each wrong answer. The test had 10 questions. Maria got 8 correct answers.

1. How many points did she get for correct answers? (5 x 8 = 40)

2. How many points were subtracted for wrong answers? (2 x 3 = 6)

3. What was her final score? (40 − 6 = 34)

1. (a) 25	(b) 20	(c) 35
2. (a) 3	(b) 5	(c) 1
3. (a) 5	(b) 45	(c) 15
4. (a) 4	(b) 6	(c) 9
5. (a) 8	(b) 10	(c) 7

6. 1 kg of prawns: $8
   5 kg of prawns: $8 x 5 = $40
   She paid **$40**.

7. 1 cake: $7
   5 cakes: $7 x 5 = $35
   She received **$35**

8. 5 people: $45
   1 person: $45 ÷ 5 = $9
   Each person spent **$9**.

9. The cookies are being grouped by 5.
   How many 5's in 25? 25 ÷ 5 = 5
   She got **5** packets.

10. 1 kg: $5
    3 kg: $5 x 3 = $15
    He paid **$15**.

11. 5 kg: $30
    1 kg: $30 ÷ 5 = $6
    1 kg of cherries cost **$6**.

Workbook

Exercise 5, p. 43

1. 15
   20
   25
   30
   35
   40
   45
   50

Exercise 6, pp. 44-45

1. 7 x 5 → 35   20 ← 5 x 4
   5 x 3 → 15   50 ← 10 x 5
   8 x 5 → 40   45 ← 5 x 9
   5 x 6 → 30   5 ← 1 x 5
   2 x 5 → 10   25 ← 5 x 5

2. 1 box: 6 cakes
   5 boxes: 6 x 5 = 30
   He bought **30** cakes.

3. 1 pot: 5 packets
   3 pots: 5 x 3 = 15
   She used **15** packets of tea.

4. 1 week: $10
   4 weeks: $10 x 4 = $40
   He spent **$40**.

Exercise 7, p. 46-48

1. 1   2
   3   7, 7
   5, 5   9, 9
   4, 4   6, 6
   8, 8   10, 10

2. clockwise from pole:
   8 → 40 ÷ 5
   7 → 35 ÷ 5
   9 → 45 ÷ 5
   4 → 20 ÷ 5
   10 → 50 ÷ 5
   5 → 25 ÷ 5
   3 → 15 ÷ 5
   6 → 30 ÷ 5

3. How many 5's in 40?
   40 ÷ 5 = 8
   There were **8** pencils in each bundle.

4. How many 5's in 50?
   50 ÷ 5 = 10
   It took 10 weeks to save **$50**.

5. 5 students: $20
   1 student: $20 ÷ 5 = $4
   Each student received **$4**.

Chapter 3 – Multiplying and Dividing by 10

Objectives

- Count by 10's.
- Relate the associated facts 10 x _____ and _____ x 10.
- Relate division by 10 to multiplication by 10.
- Solve word problems involving multiplication or division by 2, 3, 4, 5, or 10.

Notes

In this chapter your student will learn the multiplication and division facts for 10 and solve problems involving multiplication and division by 10 as well as by 2, 3, 4, and 5.

Students already learned to count by 10's in earlier levels of *Primary Mathematics*. Multiplication and division facts for 10 should be easy.

Your student will notice that to multiply a number by 10, we simply add a 0. To divide a number (that is divisible by 10) by 10 we simply remove a 0. This concept will be explored in more detail in *Primary Mathematics* 3.

Continue to have your student practice all the multiplication and division facts learned so far.

Material

- 1-50 number board (Appendix p. a18)
- Counters
- Multilink cubes
- Hundred-chart
- Number cubes
- 4 sets of number cards 1-10
- Appendix pp. a25-a26
- Mental Math 30-33 (Appendix)

(1) Multiply by 10

Discussion

Concept page 34

The eggs are being counted in groups of 10. Ask your student how many eggs she would get if she bought a number of trays up to 10. "How many eggs would you get if you bought 4 trays?"

Activity

Have your student fill in the multiplication tables for 10 (appendix p. a25). Ask him to observe any patterns. The answer for a number multiplied by 10 is simply the ten of that number.

Practice

Tasks 1-4, pp. 34-35

Ask your student to write equations for task 2.

1. (a) 40	
40	
(b) 60	
60	
2. (a) 10¢ x 7 = 70¢	
(b) $10 x 8 = $80	
3. 40	70
40	70
4. (a) 30 (b) 100 (c) 90	
(d) 20 (e) 10 (f) 60	

Discussion

Practice C, problems 6, 8, and 10, p. 36

Workbook

Exercise 8, pp. 49-51 (Answers p. 46)

Reinforcement

Continue fact practice activities, now including the multiplication facts for 10.

Mental Math 30

6. 1 ticket: $7
10 tickets: 10 x $7 = $70
She paid **$70**.

8. 1 jar: $3
10 jars: $3 x 10 = $30
She received **$30**.

10. 1 bag: 5 kg
10 bags: 5 kg x 10 = 50 kg
She bought **50 kg** of sugar.

(2) Divide by 10

Activity

Have your student fill out a copy of the chart on appendix p. a26. To divide a number by 10, we can think of what number times 10 gives that number. For example, to divide 70 by 10, think of what number times 10 is 70.

Ask your student to look at the division answers and compare them to the number we are dividing by 10. Point out that if a number can be evenly divided by 10, it will end in 0. So it is easy to find the answer; we simply make the ten we are dividing by 10 into to a one.

$70 \div 10 =$ ___

___ $\times 10 = 70$

$$7 \xrightarrow[\div 10]{\times 10} 70$$

$10 \div 10 = 1$
$20 \div 10 = 2$
$30 \div 10 = 3$
$40 \div 10 = 4$
$50 \div 10 = 5$
$60 \div 10 = 6$
$70 \div 10 = 7$
$80 \div 10 = 8$
$90 \div 10 = 9$
$100 \div 10 = 10$

Practice

Tasks 5-6, p. 35

5. 5	8, 8	
6. (a) 6	(b) 3	(c) 1
(d) 4	(e) 10	(f) 9

Discussion

Practice C, problems 7, 9, and 11, p. 36

Workbook

Exercise 9, pp. 52-54 (Answers p. 46)

Reinforcement

Continue to provide your student with opportunities to learn the math facts for multiplication and division by 2, 3, 4, and 5.

Mental Math 31

7. Group by tens.
$40 \div 10 = 4$
Emily bought **4** packets.

9. 10 cans: $80
1 can: $80 \div 10 = $8
1 can costs **$8**.

11. How many 10's in 60?
$60 \div 10 = 6$
She bought **6** boxes.

Enrichment

Discuss the following problem with your student.

⇒ 10 boxes of 8 cookies were shared among 10 boys.

1. How many cookies did each boy get? ($10 \times 8 = 80$, $80 \div 10 = 8$)

2. How many cookies altogether did 5 of the boys get? ($8 \times 5 = 40$)

3. If 5 boxes were dropped in a puddle and the cookies could not be eaten, how many cookies would each of the 10 boys get? (5 boxes left, $5 \times 8 = 40$, $40 \div 10 = 4$, or, since half the boxes are gone, each boy gets half as many, or 4 cookies.)

Practice

Practice C, p. 36, problems 1-5

(answers to 6-11 are in previous two lessons).

1. (a) 40	(b) 10	(c) 70
2. (a) 6	(b) 2	(c) 7
3. (a) 60	(b) 50	(c) 100
4. (a) 3	(b) 1	(c) 9
5. (a) 10	(b) 8	(c) 5

Practice D, p. 37

Reinforcement

Extra Practice, Unit 8, Exercise 3, pp. 121-122

Mental Math 32-33

Test

Tests, Unit 8, Chapter 3, A and B, pp. 29-32

Enrichment

Write the following problem down and see if your student can determine what steps are needed to solve it. If she cannot, guide her in determining what she needs to find for each step.

⇒ A cook made 24 cookies and 20 brownies. Somebody ate 4 cookies and 5 brownies while they were cooling. The cook put 4 cookies and 3 brownies on each plate. How many plates did she use?

1. Find the number of cookies left.
   $24 - 4 = 20$
   There are 20 cookies left.

2. Find the number of brownies left.
   $20 - 5 = 15$
   15 brownies are left.

3. Find the number of plates needed.
   $20 \div 4 = 5$
   or $15 \div 3 = 5$

1. (a) 24	(b) 20	(c) 30
2. (a) 3	(b) 4	(c) 4
3. (a) 32	(b) 25	(c) 36
4. (a) 8	(b) 7	(c) 7
5. (a) 45	(b) 30	(c) 80

6. 4 m of cloth: $36
   1 m of cloth: $36 \div 4 = 9
   1 m of cloth costs **$9**.

7. 1 dress: 3 m
   5 dresses: $3 \times 5 = 15$ m
   He used **15 m** of cloth.

8. How many $10's in $50?
   $50 \div 10 = 5$
   He can buy **5** puzzles.

9. How many 4's in 24?
   $24 \div 4 = 6$
   There are **6** chairs in each row.

10. How many 5's in 40?
    $40 \div 5 = 8$
    There were **8** boxes of pies.

11. 1 box: 4 cards
    8 boxes: $4 \times 8 = 32$ cards
    There were **32** cards.

Workbook

Exercise 8, pp. 49-51

1. 30
   40
   50
   60
   70
   80
   90
   100

2. 50   20   40
       30   32
     40   20   12
       35   50

3. 1 row: 5 soldiers
   10 rows: 5 x 10 = 50
   There are **50** soldiers in 10 rows.

4. 1 dictionary: $10
   10 dictionaries: $10 x 10 = $100
   He received **$100**.

5. 1 m of cloth: $7
   10 m of cloth: $7 x 10 = $70
   She paid **$70**.

Exercise 9, pp. 52-54

1.   3
      5   5
      6   6
      7   7
      1   1
      3   3
      8   8
      4   4
      2   2
      9   9

2.

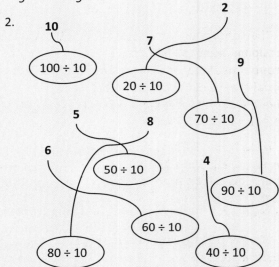

3. 10 bags: 60 kg
   1 bag: 60 kg ÷ 10 = 6 kg
   One bag weighs **6 kg**.

4. 10 plants: $40
   1 plant: $40 ÷ 10 = $4
   One plant costs **$4**.

5. 90 chairs are grouped by 10's
   90 ÷ 10 = 9
   There were **9** chairs in each row.

Chapter 4 – Division with Remainder

Objectives

♦ Use repeated subtraction to find a remainder.
♦ Relate repeated subtraction to division by grouping.
♦ Find the remainder in division.

Vocabulary

♦ Left over
♦ Remaining

Notes

In this chapter your student will learn about situations where division results in a remainder.

The concept of remainder is first introduced by looking at repeated subtraction. The process of grouping items to divide can also be represented by subtracting the quantity that goes in each group repeatedly until there are none left. For example, we can represent the process of making equal groups of 5 from 20 objects with the equation $20 - 5 - 5 - 5 - 5 = 0$. This does not, however, give the answer to the division expression, $20 \div 5$. We still need to count the groups, or the number of times we took 5 away.

If we have 23 items and want to know how many are left over after making groups of 5, we can write the equation $23 - 5 - 5 - 5 - 5 = 3$. We are again making equal groups, but there are some left over. This can be represented with the equation $23 \div 5 = 4$ R 3, where R stands for the remainder. At this level, students will not be using R to stand for the remainder. Instead, they will use the term "left over."

Division also represents sharing. If we have 23 items and want to put them into 5 groups, we could write $20 \div 5 = 4$ R 3 to show there are 4 in each group with 3 left over. In this situation the process of making 5 groups and finding how many go in each group cannot really be represented by repeated subtraction. Your student can either act out the problem, or solve it by using division facts. He needs to think of the number times 5 that gives an answer closest to 23 without being more than 23. Then he can subtract that answer from 23 to get the remainder. $4 \times 5 = 20$, $23 - 20 = 3$, which is how many are left over.

This chapter is just an introduction to remainders. Remainders will be covered more thoroughly in *Primary Mathematics* 3. Allow your student to use counters, pictures, or repeated subtraction to find the answer to problems involving remainders if needed.

Material

♦ Counters
♦ Number cube

(1) Find the remainder in division

Activity

Give your student 20 counters. Ask him to put them into groups of 5 and write an equation. He will probably write a division equation. Tell him that each time he is making a group of 5, he is taking away 5. We could show taking away 5 several times with subtraction. If we take 5 away 4 times, how much is left? None. The subtraction problem won't give the answer to how many groups there are, but it does tell us that there are none left over.

Now give your student 23 counters and ask her to put them in groups of 5. Make up a story, such as, "There are 23 donuts that have to be packed into boxes, each of which can hold 5 donuts. How many will be left over?" Tell her that we can write a subtraction equation to show how many are left over. If we also want to show how many groups there are, we can write a division equation. Since not all the donuts or counters go into equal groups, we also have to include in the answer how many are left over.

Ask your student what is the least number of boxes needed for the donuts. We will need 5 boxes in all, 4 will have 5 donuts in each box, and 1 will have the left-over 3.

Give your student 23 counters again and tell her that we want to share 23 cookies among 5 children. Since we are sharing, we can write a division equation. Write the expression 23 ÷ 5. Ask him if we will have left-overs. Since the number we are dividing by 5 does not end in 5 or 0, we know that there will be left-overs. Get him to share out the counters into 5 groups to see that there are 3 left over. Remind him that we can find the answer to a division problem using multiplication. In this case, we would ask, "What times 5 is 23?" There is no number times 5 that is 23, but we know that 4 x 5 is 20. So only 20 can be shared evenly. Each child gets 4 cookies, and there are 3 cookies left over.

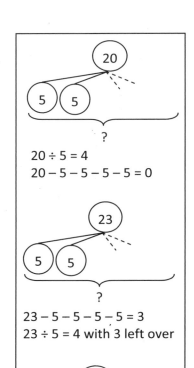

$20 \div 5 = 4$
$20 - 5 - 5 - 5 - 5 = 0$

$23 - 5 - 5 - 5 - 5 = 3$
$23 \div 5 = 4$ with 3 left over

$23 \div 5 = ?$
$4 \times 5 = 20 \quad 5 \times 5 = 25$
5 in each group is too many.
4 in each group, with 3 left over

Discussion

Concept page 38
Tasks 1-8, pp. 38-40

Allow your student to use counters with any of these problems as needed.

Workbook

Exercise 10, pp. 55-56 (Answers p. 51)

Allow your student to use counters, drawings, or repeated subtraction. Do not require a written equation for this exercise.

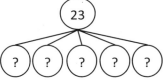

$9 - 4 - 4 = 1$

1. 2 with 1 left over
2. 4 with 2 left over
3. 2 with 1 left over
4. 5 with 6 left over
5. 2 remaining
6. 5 with 3 left over
7. 7 with 1 left over
8. 3 with 2 left over
   4 bags needed

(2) Practice

Activity

Use a number cube labeled with 1, 2, 3, 4, 5, and 10. Have your student throw the number cube and then you call out a number that is less than 10 times the number on the cube. Have him answer yes or no if there will be a remainder when that number is divided by the number on the cube. For example, he throws a 4, you call out 33, and he tells you that the remainder is 1 (4 x 8 = 32, 33 − 32 = 1). Allow your student to use counters or repeated subtraction.

Practice

Practice E, p. 41

Allow your student to use repeated subtraction, manipulatives, or drawings to solve these if needed.

Reinforcement

Extra Practice, Unit 8, Exercise 4, pp. 123-124

Test

Tests, Unit 8, Chapter 4, A and B, pp. 33-38

Enrichment

Give your student some division problems (for 2, 3, 4, 5, or 10 only) out loud, including ones that give a remainder, and have her find the answer and the remainder, if any.

---

1. 23 − 10 − 10 = 3
   Or 23 ÷ 10 = 2 with 3 left over
   There were **3** soldiers left over.

2. 13 ÷ 3 = 4 with 1 left over
   1 dog was not in a group.

3. 37 ÷ 10 = 3 with 7 left over.
   She used **3** vases.
   7 orchids were left over.

4. 25 − 7 − 7 − 7 = 4
   (a) He made **3** groups of 7.
   (b) He had **4** almonds left.

5. 26 ÷ 10 = 2 with 6 left over.
   3 vans are needed.

6. 38 ÷ 5 = 7 with 3 left over.
   8 boxes are needed.

Review 8

Review

Review 8, pp. 42-43

You can do this review all at once, or you can select problems each day for more continuous review as you continue in the textbook.

Workbook

Review 9, pp. 57-60

Reinforcement

Mental Math 34-35

Continue to help your student practice math facts.

Test

Tests, Units 1-8, Cumulative, A and B, pp. 39-45

1. (a) 12   (b) 40   (c) 90

2. (a) 5   (b) 5   (c) 5

3. (a) 36   (b) 70   (c) 16

4. (a) 6   (b) 8   (c) 1

5. (a) 100   (b) 28   (c) 15

6. 5 boxes: 50 cupcakes
   1 box: 50 ÷ 5 = 10 cupcakes
   There were **10** cupcakes in each box.

7. 1 bag: 10 oranges
   6 bags: 10 x 6 = 60 oranges.
   She bought **60** oranges.

8. How many $4's in $40?
   $40 ÷ $4 for each kg = 10 kg
   He bought **10 kg** of lychees.

9. 2 boys: $16
   1 boy: $16 ÷ 2 = $8
   Each boy received **$8**.

10. 39 ÷ 4 = 9 with 1 left over.
    They need **10** tents.

11. (a) Group 24 m into 4 m.   24 ÷ 4 = 6
        She made **6** sets of curtains.
    (b) There was no material left over.

12. (a) 800   (b) 648   (c) 902

13. (a) 260   (b) 9   (c) 401

14. (a) 12   (b) 30   (c) 70

15. (a) 10   (b) 6   (c) 9

16. (a) **7**
    (b) 15 − 7 = 8
        The apple weighs **8** oz.
    (c) The apple weighs **1 oz** more.

17. 9 in. − 5 in. = 4 in.
    The other piece is **4 in.** long.

18. 1 lb: $3
    7 lb: $3 x 7 = $21
    She paid **$21**.

19. 27 yd ÷ 5 = 5 yd with 2 yd left over
    (a) She made **5** dresses.
    (b) She had **2 yd** of cloth left over.

Workbook

Exercise 10, pp. 55-56

1. 4
   3
   5 bags are needed.

2. (a) 27 − 7 = 20
      She shared 20 brownies
      20 ÷ 4 = 5
      Each friend gets **5** brownies.
   (b) 7 − 4 = 3
      There will be **3** brownies left.

3. 9 ÷ 2 = ?
   4 x 2 = 8, so
   9 ÷ 2 = 4 with 9 − 8 = 1 left over
   (a) There are **4** balls in each group.
   (b) **Yes**, there is **1** ball left over.

4. 26 ÷ 6 = ?
   4 x 6 = 24, so
   26 ÷ 6 = 4 with 26 − 24 = 2 left over
   2 beads are left over.

5. 37 ÷ 4 = ?
   9 x 4 = 36, so
   37 ÷ 4 = 9 with 37 − 36 = 1
   1 in. of ribbon is left over.

Review 9, pp. 57-60

1. (a) 336   (b) 84
   (c) 584   (d) 206
   (e) 798   (f) 302

2. (a) 44; 300   (b) 22; 40
      344   62
       344   62

3. (A) 12   14   5   8
   (B) 27   12   6   8
   (C) 32   24   5   7
   (D) 25   45   6   8
   (E) 30   50   6   9
   (F) 30   10   10   10
   (G) 18   9   7   9

4. 402 − 382 = 20
   She sold **20** tickets the second day.

5. 1 box: 10 pencils
   3 boxes: 10 x 3 = 30 pencils
   There were **30** pencils.

6. Make groups of 5.
   43 ÷ 5 = ?
   8 x 5 = 40, so
   43 ÷ 5 = 8 with 43 − 40 = 3 left over.
   9 boats are needed.

7. 1 kg: $5
   9 kg: $5 x 9 = $45
   9 kg of peaches cost **$45**.

8. 122 − 86 = 36
   He needs **36** envelopes.

9. 386 + 255 + 145 = 786
   There were **786** people.

Unit 9 – Money

Chapter 1 – Dollars and Cents

Objectives

- Count money in a set of bills and coins up to $100.00.
- Recognize, read, and write the decimal notation for money.
- Make equivalent values of money using different denominations.
- Convert cents to dollars and cents.
- Convert dollars and cents to cents.
- Make change for $1, $5, and $10.
- Mentally subtract cents from $1.
- Mentally subtract dollars and cents from $10.

Notes

In *Primary Mathematics* 1B students learned to recognize and compare coins and bills, to recognize the symbols $ and ¢, and to count money in a set of coins up to $1 or a set of bills up to $10.

In this chapter your student will learn to count money in sets of bills and coins up to $100, to convert from dollars and cents to cents and vice-versa, and to make change for $1, $5, and $10.

The concept of decimals has not yet been taught. The decimal point should be presented as a dot separating dollars from cents. Decimals will be taught formally in *Primary Mathematics* 4B.

To make change for $1, your student can count up with coins, generally starting with the smallest denomination. He can practice making change in different ways. To subtract from $1, he can use the mental math skills for making 100 learned in Unit 7.

To make change for $10, your student can count up coins to the next dollar, and then count up dollars to $10. To subtract from $10, she can subtract the dollars from $9, and the cents from 100¢ using mental math strategies.

The second lesson of this chapter and its accompanying exercise focuses on reading the money amounts in words. Being able to actually write the amounts in words, spelling the number words correctly, is a necessary skill, but could be taught in a spelling lesson rather than a math lesson. Also, your student's skill in writing and spelling may not be at the same stage as his skill in math or reading the number words. So use your own discretion on how much time you want to have him spend writing the money amounts.

Material

- Coins
- Bills ($1, $5, $10, $20)
- Store cards (see explanation of manipulatives in the introduction to this guide)
- Mental Math 36-38 (Appendix)

(1) Write money in dollars and cents

Activity

Give your student the following amounts of money and ask him to count the money.

1. First 1, and then 2 half-dollars, if you have them.

2. 1 quarter, then 2, 3, and 4 quarters. Remind your student that 100 cents is also 1 dollar.

3. Different amounts of dimes and nickels up to a dollar.

4. 1 quarter and some nickels, then 1 quarter and some dimes, and then 1 quarter, a nickel, and some dimes. Point out that it is easy to count on the 5 cents for the nickel before counting on by tens for the dimes.

5. A set of coins up to a dollar, including some of each denomination. Point out strategies for counting the amount of money, such as starting with the highest denomination, or combining a quarter with a nickel. Write the amounts as cents, e.g. **67¢**. Remind your student that ¢ is the symbol for cents.

6. A set of bills. Point out that it is easier to count the money if we make groups of the same denomination and start with the largest. Write the amounts as dollars, e.g. **$32**. Remind your student that $ is the symbol for dollar, and we put it in front of the money, rather than after, as with cents.

Discussion

Concept pages 44-45

Point out that when we have both dollars and cents, we write a dollar sign, the total number of dollars, a dot, and then the total number of cents. We do not write a cent sign after the cents if we are using a dollar sign and a dot for money amounts.

Tasks 1-2, pp. 45-46.

Ask your student to write the amounts for task 1 in figures and say the amounts for both tasks out loud.

| 1. (a) $23.30 |
| (b) $4.32 |
| (c) $8 or $8.00 |
| (d) $0.65 |

Activity

Give your student the following amounts of money and ask her to count and write the amounts in dollars.

1. A set of bills and coins where there are less than 10 cents in coins, e.g. $3.06. Tell your student that if there are less than 10 cents, we write a 0 for the tens after the dot separating dollars and cents. We always have 2 numbers for the cents. The first place after the dot is always for the tens of the amount of cents, and the second place for the ones. Write $3.60 and ask your student how that is different from $3.06.

2. A set of bills and coins where the coins add up to $1 exactly, e.g. $23. Tell your student that even though there are cents, they equal a dollar, and we can write the total amount as the dollar amount, or include the dot and two zeros for the cents.

Workbook

Exercise 1, pp. 61-63 (Answers p. 60)

(2) Read and write money amounts

Discussion

Task 3, p. 46

After your student has supplied the numbers, tell him that sometimes we need to write the amount out in words, such as on a check, or in legal documents. Refer back to the amount in words at the top of p. 45 in the textbook.

3. (a) 4	75
(b) 8	0
(c) 0	35

Guide your student in writing out in words the amounts in tasks 2 and 3. Point out that we write "dollars" after the dollar amounts, the word "and" before the cent amount and then the word "cents" even though the amounts in the book are written only with a dollar sign. Although we can write simply the number of dollars if there are no cents, for this lesson get your student to write "and zero cents" when the amount shows 00 for cents. Eventually, when they write amounts on checks, they will have to include cents, even though it is written as a fraction on checks (e.g. thirty-eight dollars and 00/100 cents). Have her also write "zero dollars" if there are no dollars.

2. eight dollars and fifty cents
zero dollars and sixty cents
seventeen dollars and ninety-five cents
five dollars and zero cents
3. (a) four dollars and seventy-five cents
(b) eight dollars and zero cents
(c) zero dollars and thirty-five cents

Give your student as much additional practice as you feel necessary in reading and writing money amounts.

Workbook

Exercise 2, pp. 64-67 (Answers p. 60)

Reinforcement

The following suggestions can be applied to reinforce the previous lesson. If you want to reinforce reading and writing money amounts as well, modify them to include writing out the amounts in words for your student to read or asking her to write the amounts in words instead of just figures.

Give your student sets of money to count and write the amount in figures. Do not include sets of money where the coins add up to more than $1 until after the next lesson.

Provide a set of coins and bills. Write an amount of money, e.g. $10.43 (in figures or in words), and have your student count out the correct amount of coins and bills.

(3) Change from one denomination to another

Discussion

Task 4, p. 46, and 6(a), p. 47

Ask your student for other ways that coins can be changed for other kinds of coins, such as 5 nickels, or two dimes and a nickel for a quarter.

Task 6(b) and 7, p. 47

Tell your student that the most common bills are $1, $5, $10, and $20, but there are also fifty-dollar bills and hundred-dollar bills. Ask him how many twenty dollar bills can be changed for a hundred-dollar bills, or how many ten dollar bills can be changed for a fifty-dollar bill.

4. (a) 100
(b) 10
(c) 20
6. (a) 4
6. (b) 4
7. (a) $23

Activity

Give your student a set of coins that total more than $1 and have her count the money. Write down the total amount in cents, e.g. 152¢. Ask her how many cents are in a dollar. Write 100¢ = $1 = $1.00. Have her set aside $1 of the coins. Trade it in for a dollar bill and ask how much money there is. Write 152¢ = $1.52. Tell her that even if the money is all in coins, 100¢ is still a dollar, and we can write the amount as dollars and cents.

Write an amount of money that is more than $2 in cents, e.g. 343¢. and ask your student to write the total amount in dollars. Do the same with an amount that is less than 10¢.

Write an amount of money a little over $1, such as $1.32, and have your student count out the money with just coins. Tell him he has counted out a dollar and thirty-two cents, but since a dollar is 100 cents, he has a total of 132 cents. Write $1.32 = 132¢. Replace the $1 worth of coins with a dollar bill. Tell him that this amount is still the same as 132¢, even though there is a dollar bill. When we write the amount of money, we do not have to use a dollar sign if the amount has dollar bills, or the cent sign if the amount only has coins. Both ways of writing the amount of money show the same amount of money.

152¢
100¢ = $1 = $1.00
152¢ = $1.52
343¢ = $3.43
8¢ = $0.08
$1.32 = 132¢

Discussion

Task 5, p. 46

Ask your student to write the amounts in cents as well as in dollars.

Practice

Tasks 8-9, p. 47

Workbook

Exercise 3, pp. 68-69 (Answers p. 60)

5. (a) $1.50, 150¢
(b) $1.50, 150¢
8. (a) 0.65
(b) 1.65
9. (a) 85
(b) 120
(c) 200
(d) 205

Reinforcement

Extra Practice, Unit 9, Exercise 1, pp. 129-132

Mental Math 36

Enrichment

Refer to task 7, p. 47. Ask your student for a different way that Melissa could have had $23. You can record your student's suggestions in a table like the one here.

$20	$10	$5	$1
1			3
	2		3
	1	2	3
	1	1	8
	1		13
		4	3
		3	8
		2	13
		1	18
			23

Ask your student to find all different combinations of coins that total 10¢ and record her answers in a chart.

You can show him a systematic way of finding all the different possibilities. Start with a dime. That is one possibility. Then, trade in the dime for the next smaller denomination, which is nickels. Two nickels is another possibility. Then, trade in 1 of the nickels for the next lower denomination (pennies) and finally the last nickel for pennies.

10¢	5¢	1¢
1		
	2	
	1	5
		10

Draw a table with the headings shown at the right and provide your student with coins. Ask him to find all the possible ways to make 25¢. There are 13 possible combinations.

You can have your student find different combinations for other amounts, but keep the amount small and do not require all combinations if there are too many. (There are 292 different possible coin combinations for making $1 from pennies, dimes, nickels, quarters, and half-dollars).

quarter	dime	nickel	penny
1			
	2	1	
	2		5
	1	3	
	1	2	5
	1	1	10
	1		15
		5	
		4	5
		3	10
		2	15
		1	20
			25

Game

Material: Coins.

Procedure: One player picks out a set of coins less that $1 but does not let the other players see it. He tells the other players the total amount and the number of coins. The other players try to guess what coins he has. They can use coins to help them come up with possible answers.

(4) Make change for $1

Activity

Give your student some coins. Pretend that you want to buy some item for less than $1 and have her give you change for $1. Discuss strategies for counting up as she counts out the correct change. For example, if the item is 43 cents, she can count out 2 pennies to get to 45 cents, and then count up 5 to 50, as she hands you a nickel, and then count up 25 to 75 cents and then 1 dollar as she gives you 2 quarters. You can use a restriction that you get the least amount of coins possible, so she can't give you 2 pennies, a nickel, and 5 dimes.

Repeat, but this time remove all coins of one denomination (except pennies) from the money your student has to make change with. So, for example, he has to make change when he does not have any dimes.

Discussion

Task 10, p. 48

10. 55

Ask your student to find the change by subtracting 45 cents from $1. Since $1 is 100 cents, he can use mental math strategies for subtracting from 100.

Practice

Tasks 11-12, p. 48

Workbook

Exercise 4, p. 70 (Answers p. 60)

11. (a) 40¢   (b) 15¢
(c) 90¢   (d) 95¢
12. (a) 80¢   (b) 25¢

Reinforcement

Mental Math 37

(5) Make change for $5 and $10

Activity

Give your student some coins and bills and pretend you want to buy something for under $5 or under $10. Give him a 5-dollar or 10-dollar bill and have him give you change. Discuss strategies for making change. Generally, he should count out coins up to the next dollar, and then bills to $5 or $10.

Write the expression $10 − $6.35 and discuss strategies to solve it.

We can add cents up to the next dollar, and then dollars up to $10.

We can think of the $10 as $9 and 100¢, and then subtract $6 from $9 and 35¢ from 100¢.

Give your student some other problems to solve. You can include subtraction from any dollar amount with $10, since she can subtract dollars from one less dollar and cents from 100¢.

> $10 − $6.35
>
> $6.35 $\xrightarrow{\quad +65¢ \quad}$ $7 $\xrightarrow{\quad +$3 \quad}$ $10
>
> $10 $\Big\langle$ $9 − $6 = $3
> $\qquad\qquad$ 100¢ − 35¢ = 65¢
>
> $10 − $6.35 = $3.65

⇒　$10 − $3.15　　　　　　　　　($6.85)

⇒　$10 − $6.05　　　　　　　　　($3.95)

⇒　$10 − $2.42　　　　　　　　　($7.58)

⇒　$10 − $9.91　　　　　　　　　($0.09)

⇒　$5 − $1.85　　　　　　　　　($3.15)

⇒　$7 − $3.36　　　　　　　　　($3.64)

Practice

Tasks 13-14, p. 48

Workbook

Exercise 5, problems 1-2, p. 71 (Answers p. 60)

Reinforcement

Mental Math 38

13. (a) $5.70		(b) $7.35
14. (a) $4.60		(b) $3.05

(6) Solve Word Problems

Practice

Practice A, p. 49

Have your student do problems 1-4 and discuss problems 5 and 6 with him. Both are subtraction problems. In 5, the total is the $1 given to pay for the file and the missing part is the change. In 6, the total is the cost of the teddy bear. One part is how much she has and the other how much she still needs.

Workbook

Exercise 5, problems 3-5, p. 71-74

Reinforcement

Play store. Tag some items with a cost. Your student can play cashier, or can select items to buy. If she is the buyer, give her a specific amount of money under $10. She will have to estimate to select items that do not exceed how much money she has.

Continue to help your student to work on multiplication and division facts for 2, 3, 4, 5, and 10.

Game

<u>Material</u>: Store cards, bills and coins.

<u>Procedure</u>: Shuffle cards and place face-down. Give each player four $10 bills (or index cards with $10 written on them to represent $10 bills), and put the coins and smaller bills in the center of the playing area. Each player takes turns turning over a card and buying the item, putting money in the center and retrieving change. The first player to use up all his money or not have enough to buy the last item wins.

Test

Tests, Unit 9, Chapter 1, A and B, pp. 47-50

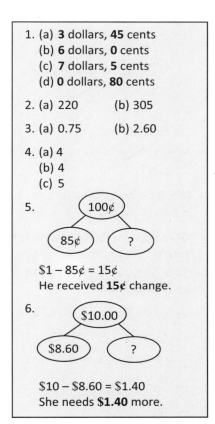

1. (a) **3** dollars, **45** cents
   (b) **6** dollars, **0** cents
   (c) **7** dollars, **5** cents
   (d) **0** dollars, **80** cents

2. (a) 220   (b) 305

3. (a) 0.75   (b) 2.60

4. (a) 4
   (b) 4
   (c) 5

5.
   100¢
   85¢   ?

   $1 − 85¢ = 15¢
   He received **15¢** change.

6.
   $10.00
   $8.60   ?

   $10 − $8.60 = $1.40
   She needs **$1.40** more.

Workbook

Exercise 1, pp. 61-63

1. → $0.95
   → $0.59
   → $1.65
   → $1.56
   → $2.25

2. (a) $0.92
   (b) $3.20
   (c) $5.85
   (d) $6.04
   (e) $18.05

3. (a) $0.84
   (b) $24.00
   (c) $58.40
   (d) $58.55

Exercise 2, pp. 64-67

1. left side: right side:
   → 5 dollars 45 cents   9 dollars 60 cents ←
   → 5 dollars 50 cents   8 dollars   ←
   → 4 dollars 40 cents   6 dollars 90 cents ←
   → 85 cents   4 dollars 5 cents ←

2. $3.05
   $4.30
   $5.00
   $0.50 or 50¢
   $9.75
   $9.90

3. **6** dollars **80** cents
   4 dollars **65** cents
   0 dollars **70** cents
   6 dollars **45** cents
   7 dollars **0** cents

4. → $23.00
   → $4.00
   → $13.30
   → $0.20
   → $7.50
   → $99.05

5. $0.15; $20.00; $47.00; $74.50; $30.45;
   $86.05; $47.15; $0.95; $95.05; $40.25

Exercise 3, pp. 68-69

1. → 180¢
   → 270¢
   → 345¢
   → 105¢
   → $0.10
   → $0.05
   → $0.35
   → $3.00

2. $1.00 $2.05
   $2.00 $1.90
   $1.25 $3.50
   $2.40 $0.85
   $3.60 $0.70
   $4.05 $0.05

3. $0.30   10¢
   $0.45   75¢
   $1.20   105¢
   $2.50   305¢
   $3.00   250¢
   $0.75   150¢
   $3.45   400¢
   $0.06   8¢

Exercise 4, p. 70

1. (a) $0.55   (b) $0.05
   (c) $0.25   (d) $0.65

2. clockwise from lower left
   $0.15; $0.25; $0.20; $0.30; $0.35; $0.45

Exercise 5, pp. 71-74

1. (a) $0.80   (b) $1.60
   (c) $7.40   (d) $6.90

2. (a) 50; 3.00; 3.50
   (b) 85; 7.00; 7.85
   (c) 5.30

3. Change = $1 − 45¢ = **55¢**

4. Money left = $10 − $5.20 = **$4.80**

5. (a) 55¢ + 45¢ = **$1.00**
   (b) $1 − 85¢ = **15¢**
   (c) $10 − $8.20 = **$1.80**
   (d) $10 − $4.40 = **$5.60**
   (e) table-tennis racket and flying saucer

Chapter 2 – Adding Money

Objectives

♦ Add dollars to dollars and cents.
♦ Add cents to dollars and cents.
♦ Add cents to make the next dollar.
♦ Add money within $10 by first adding dollars, and then adding cents.
♦ Add money within $10 using the addition algorithm.
♦ Add money with a cent amount close to $1 by adding $1, and then subtracting the difference.
♦ Solve word problems involving the addition of money.

Notes

In this chapter your student will learn the following strategies for adding money within $10.

⇒ Add the dollars, and then add the cents.

$4.15 + $3.50 = ?
$4.15 $\xrightarrow{\ +\$3\ }$ $7.15 $\xrightarrow{\ +50¢\ }$ $7.65
$4.15 + $3.50 = $7.65

⇒ Use the formal algorithm for addition.
Write the problem vertically, aligning the dots (decimals) and add using the same methods as with whole numbers. This method should be used when the problem cannot be easily solved mentally.

$$
\begin{array}{r}
\$4.\,85 \\
+\ \ \$3.\,55 \\
\hline
\$8.\,40
\end{array}
\quad\leftarrow\quad
\begin{array}{r}
1\ \ 1 \\
4\ 8\ 5 \\
+\ 3\ 5\ 5 \\
\hline
8\ 4\ 0
\end{array}
$$

⇒ (Optional) Add cents by first making a whole number of dollars. This method can be used when the cents add to more than a dollar, particularly when it is easy to see what needs to be added to one set of money to make a whole dollar, and what remains when this amount is subtracted from the other set of money.

$7.25 + $0.85 = $8 + $0.10 = $8.10
$\wedge$
75¢ 10¢

⇒ Add a whole number of dollars, and then subtract the difference. This method can be used when the cents in one of the amounts being added are close to 100.

$6.25 + $2.95 = ?
$6.25 $\xrightarrow{\ +\$3\ }$ $9.25 $\xrightarrow{\ -5¢\ }$ $9.20
$6.25 + $2.95 = $9.20

The problems at this level in the textbook and workbook will generally use multiples of 5 for the cents to facilitate mental calculation.

Material

♦ Bills and coins
♦ Store cards
♦ Mental Math 39-40 (Appendix)

(1) Add dollars and cents mentally

Note

This lesson should be fairly easy for students who have done previous levels of *Primary Mathematics*. If it is not, split it into two lessons and allow your student to use bills and coins to do some of the tasks if needed.

Discussion

Concept page 50

| $7.65 |

Since dollars and cents can be split into dollar amounts and cent amounts, when adding money we can add dollars to dollars and cents to cents.

Tasks 1-3, p. 51

Go through these problems with your student. She should be able to do these mentally. For task 3 she should recognize that the cents add up to a dollar. So the total dollar amount is increased by 1 dollar.

1. (a) $6.95
   (b) $14.45

2. (a) 75¢
   (b) $2.75
   (c) $5.75

3. (a) $1   (b) $3   (c) $4
   (d) $1   (e) $2   (f) $4

Tasks 4-5, p. 51

In order to add both dollars and cents, we can first add the dollars, and then add the cents.

4. (a) 6.75   6.95
            6.95
   (b) 8.65   8.80
            8.80

5. (a) $9.80   (b) $9.75
   (c) $9.90   (d) $7.80

Workbook

Exercise 6, p. 75 (Answers p. 66)

Exercise 7, p. 76 (Answers p. 66)

Reinforcement

Mental Math 39

(2) Add dollars and cents using the addition algorithm

Activity

Write the addition expression **$5.87 + $2.59**. Tell your student that we can change the dollars and cents to cents, rewrite the problem with one number under the other, and add the normal way starting with the ones. Then we put the answer back into dollars and cents. Any addition problem can be done this way, and we can use this method when it is easier to do so than trying to solve it using mental math, such as with problems where the cents add up to more than a dollar.

$5.87 + $2.59 = ?

```
    1 1
    5 8 7
+   2 5 9
    8 4 6
```

$5.87 + $2.59 = $8.46

Practice

Task 6, p. 52

6. (a) $4.25   (b) $7.55
   (c) $7.50   (d) $8.45

Workbook

Exercise 8, p. 77 (Answers p. 66)

Enrichment

Even when the cents add up to more than a dollar, it is easy to use mental math for some problems. You can use task 3 on p. 51, where the cents add up to exactly $1 and discuss a strategy similar to the "making 10" strategy your student learned in earlier levels of *Primary Mathematics*. Instead they "make $1." Write the following expressions and discuss strategies for finding the answer.

⇒ $5.65 + $1.45

Add the dollars first. Since 65 cents + 45 cents will be more than $1, add them by "making $1" or "making 100" with the cents. 65 needs 35 more to make 100. Take that from the 45 cents, leaving 10 cents. So the answer is 1 more dollar and ten cents.

$5.65 + $1.45 = $6.65 + 45¢

$1 45¢   35¢ 10¢

= $7 + $0.10
= $7.10

⇒ $6.25 + $1.85

Since 85¢ is closer to $1, it might be easier to take 15¢ from the 25¢, but if we are used to adding quarters, we could just as easily take 3 quarters from 85¢ to make the dollar.

$6.25 + $1.85 = $7.25 + 85¢

75¢ 10¢

= $8 + $0.10
= $8.10

Tell your student that this strategy works best when it is easy to quickly see what is needed to make a dollar with one set of cents, and how much will be left with the other set of cents, such as when the cents end in 0 or 5.

Have your student solve a few additional examples.

⇒ $3.90 + $0.35   ($4.25)

⇒ $4.50 + $2.65   ($7.15)

⇒ $1.65 + $3.75   ($5.40)

(3) Add dollars and cents when the cents is almost $1

Activity

Use bills and coins. Ask your student to count out $3.55 and 95¢, putting them in separate piles. Point out that 95¢ is close to $1. Ask her how much more is needed to make $1 with the 95¢; 5¢ is needed. If we take it from the $3.55, we have 5¢ less there. So we have made $1 with one set of money and subtracted 5¢ with the other set. To add 95¢ to $3.55, we can add $1 and take away 5¢.

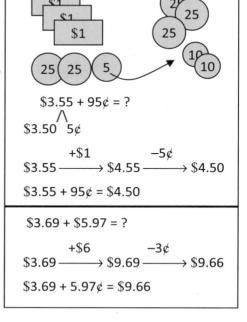

$3.55 + 95¢ = ?

$3.50   5¢

$3.55 $\xrightarrow{+\$1}$ $4.55 $\xrightarrow{-5¢}$ $4.50

$3.55 + 95¢ = $4.50

Write the problem **$3.69 + $5.97**. You can have your student show the amounts with money. Point out that 97¢ is close to $1. Ask her what is needed to make $1 with 97¢; 3¢ is needed. So to add these two money amounts, we can add $6 (the $5 in the second set plus an extra $1) and then take away 3¢.

$3.69 + $5.97 = ?

$3.69 $\xrightarrow{+\$6}$ $9.69 $\xrightarrow{-3¢}$ $9.66

$3.69 + 5.97¢ = $9.66

Discussion

Task 7, p. 52

7. (a) 7.25   7.20
            7.20
   (b) 6.60   6.59
            6.59

Practice

Task 8, p. 52

8. (a) $4.35   (b) $7.60
   (c) $6.14   (d) $6.24

Workbook

Exercise 9, p. 78 (Answers p. 66)

Reinforcement

Mental Math 40

(4) Solve word problems

Discussion

Tasks 9-10, p. 53

Make sure your student understands why the problems can be solved using addition. Allow him to choose what strategy he wants (mental math or standard algorithm).

Task 9: There are two parts, the cost of the meal and the amount of money left. The money he had at first, which we are to find, is the total.

Task 10: we need to find the total price of the stuffed toy, which is the cost of the car plus how much more it costs than the car.

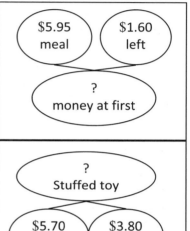

9. $7.55
   7.55

10. $9.50
    9.50

Workbook

Exercise 10, pp. 79-80 (Answers p. 66)

Problem 6 lists bills and coins. Your student can simply find the answer mentally counting the money, write down partial sums, e.g. the total of the bills, then the 8 quarters, etc., draw pictures of the money, or use actual bills and coins.

Reinforcement

Use the store cards. Shuffle and place face down. Have your student turn over two at a time, make up a story where the two amounts need to be added, and find the total.

Extra Practice, Unit 9, Exercise 2, pp. 133-136

Test

Tests, Unit 9, Chapter 2, A and B, pp. 51-54

Workbook

Exercise 6, p. 75

1. (a) 4.85
   (b) 4.45
   (c) 10.05
   (d) 13.70
   (e) 2.55
   (f) 1.90
   (g) 2.75
   (h) 3.80
   (i) 4.00
   (j) 3.00
   (k) 5.00
   (l) 5.00

Exercise 7, p. 76

1. (a) 3.45 3.75
           3.75
   (b) 5.60 5.85
           5.85
   (c) 5.15 5.80
           5.80

2. (a) $3.60
   (b) $6.90
   (c) $4.90
   (d) $4.80
   (e) $5.60
   (f) $8.90

Exercise 8, p. 79

1. E $3.05   F $5.45   L $3.55
   N $8.35   O $8.10   R $9.20
   S $6.20   U $6.00   W $9.10
   SUNFLOWER

Exercise 9, p. 78

1. (a) $3.44
   (b) $8.14
   (c) $5.54
   (d) $6.24

2. (a) $4.75
   (b) $3.60
   (c) $6.35
   (d) $8.30

Exercise 10, pp. 79-80

1. $2.40 + $3.25 = $5.65
   She spent **$5.65**.

2. $5.98 + $1.22 = $7.20
   He had **$7.20**.

3. $3.45 + 65¢ = $4.10
   She spent **$4.10**.

4. $4.50 + $2.35 = $6.85
   Her brother has $6.85.

5. $2.45 + 95¢ + $3 = $6.40
   She saved **$6.40**.

6. $20 $20
   $5 $5 $5
   $1 $1 $1 $1 $1 $1
   25¢ 25¢ 25¢ 25¢ 25¢ 25¢ 25¢ 25¢
   10¢ 10¢ 10¢ 10¢
   5¢ 5¢ 5¢
   18 pennies (10¢, 5¢, 3¢)
   $63.73 total
   He had **$63.73**.

Chapter 3 – Subtracting Money

Objectives

- Subtract dollars from dollars and cents.
- Subtract cents from dollars and cents.
- Subtract cents from a whole number of dollars.
- Subtract money within $10 by first subtracting dollars, and then subtracting cents.
- Subtract money within $10 using the subtraction algorithm.
- Subtract money with a cent amount close to $1 by subtracting $1, and then adding the difference.
- Solve word problems involving the subtraction of money.

Notes

In this chapter your student will learn the following strategies for subtracting money within $10.

$\Rightarrow$ Subtract the dollars, and then subtract the cents.

$$\$4.65 - \$3.50 = ?$$

$$\$4.65 \xrightarrow{\ -\$3\ } \$1.65 \xrightarrow{\ -50¢\ } \$1.15$$

$$\$4.65 - \$3.50 = \$1.15$$

$\Rightarrow$ Use the formal algorithm for subtraction.
Write the problem vertically, aligning the dots (decimals) and subtract using the same methods as with whole numbers. This method should be used when the problem cannot be easily solved mentally.

$$
\begin{array}{r}
\$7.\,0\,5 \\
-\ \ \$3.\,6\,8 \\
\hline
\$3.\,3\,7
\end{array}
\qquad
\begin{array}{r}
{}^{6\ 9} \\
7\,\cancel{0}^{1}5 \\
+\,3\,6\,8 \\
\hline
3\,3\,7
\end{array}
\ \leftarrow
$$

$\Rightarrow$ (Optional) Subtract cents when there are not enough by subtracting from a dollar.

$$\$5.25 - \$0.75 = \$4.25 + \$0.25$$
$$\qquad\qquad\ \ \ \ = \$4.50$$
$$\$4.25 \quad \$1$$

$\Rightarrow$ Subtract a whole number of dollars, and then add the difference. This method can be used when the cents in the amount being subtracted are close to 100.

$$\$6.25 - \$2.95 = ?$$

$$\$6.25 \xrightarrow{\ -\$3\ } \$3.25 \xrightarrow{\ +5¢\ } \$3.30$$

$$\$6.25 - \$2.95 = \$3.30$$

The problems at this level in the textbook and workbook will generally use multiples of 5 for the cents to facilitate mental calculation.

Material

- Bills and coins
- Store cards
- Mental Math 41-43 (Appendix)

(1) Subtract dollars and cents mentally

Note

This lesson should be fairly easy for students who have done previous levels of *Primary Mathematics*. If it is not, split it into two lessons, and allow your student to use bills and coins to do some of the tasks if needed.

Discussion

Concept page 54

$5.25

Since dollars and cents can be split into dollar amounts and cent amounts, when subtracting money we can subtract dollars from dollars and cents from cents.

Tasks 1-2, p. 55

Discuss the first one or two problems in each set and have your student answer the rest. He should be able to do them using mental strategies.

1. (a) $5.15 (b) $4.35 (c) $0.80
   (d) $0.45 (e) $2.45 (f) $3.45

2. (a) 60¢
   (b) $2.60
   (c) $9.60

Task 3, p. 55

(a): Subtract 90¢ from $1 by thinking of $1 as 100¢. So we make 100 with 90.

(b): To subtract 60¢ from $4, subtract the 60¢ from one of the dollars. So the dollars in the answer is one less dollar, or $3, and the cents is what we need to make 100 with 60¢.

(c)-(f): Let your student answer these on her own.

3. (a) $0.10 (b) $3.40 (c) $5.50

$$\$1 - 90¢ = 100¢ - 90¢$$
$$= 10¢$$
$$= \$0.10$$

$$\$4 - 60¢ = \$3.40$$
$$\overset{\wedge}{}$$
$$\$3\ \ 100¢$$

Tasks 4-5, p. 55

In order to subtract both dollars and cents, first subtract the dollars, and then the cents.

4. (a) 3.90　　　3.40
        3.40
   (b) 1.65　　　1.60
        1.60

5. (a) $6.20　　(b) $3.55
   (c) $0.40　　(d) $4.15

Workbook

Exercise 11, p. 81 (Answers p. 73)

Exercise 12, p. 82 (Answers p. 73)

Reinforcement

Mental Math 41

(2) Subtract dollars and cents using the subtraction algorithm

Activity

Write the subtraction expression **$7.15 – $2.59**. Tell your student we can change the dollars and cents to cents, rewrite the problem with one number under the other, and subtract starting with the ones. Then we put the answer back into dollars and cents. We can use this method when it is easier to do so than trying to solve it using mental math, such as with problems where there are not enough cents to subtract from.

$7.15 – $2.59

$$\begin{array}{r} 6\,{}^{1}0 \\ \cancel{7}\,\cancel{1}{}^{1}5 \\ -\ 2\ 5\ 9 \\ \hline 4\ 5\ 6 \end{array} \quad \text{or} \quad \begin{array}{r} 6\,{}^{1}0 \\ \$\cancel{7}.\cancel{1}{}^{1}5 \\ -\ \$2.5\ 9 \\ \hline \$4.5\ 6 \end{array}$$

$7.15 – $2.59 = $4.56

Practice

Task 6, p. 56

6. (a) $4.80   (b) $3.65
   (c) $2.65   (d) $4.30
   (e) $1.60   (f) $1.65

Workbook

Exercise 13, p. 83 (Answers p. 73)

Enrichment

Even when there are not enough cents to subtract from, sometimes it is easy to subtract the cents from one of the dollars. You can use task 3 on p. 55 where the cents are subtracted from a $1 to discuss a strategy similar to the "subtract from a ten" strategy your student learned in earlier levels of *Primary Mathematics*. Discuss the following problems.

⇒ $5 – $0.25

Because we know there are 4 quarters in a $1, it is easy to remember that taking away a quarter from a dollar leaves 3 quarters, or 75¢. So we can subtract $0.25 from one of the five dollars, leaving $4.75.

$5 – $0.25 = $4.75
∧
$4 $1

⇒ $5.10 – $0.25.

This can be solved in the same way as the last problem, except that we need to add the 10¢ to $4.75.

$5.10 – $0.25 = $4.85
∧
$4.10 $1

⇒ $7.10 – $2.25.

Subtract the dollars, and then subtract the $0.25 the same way as in the last two examples.

$7.10 – $2.25 = $5.10 – $0.25
∧
$4.10 $1
= $4.85

⇒ $8.05 – $3.50

Subtract the dollars, then 50 cents from a dollar and add back 5 cents.

$8.05 – $3.50 = $5.05 – $0.50
= $4.55

Tell your student that this strategy works best when it is easier to do when the cents end in 0 or 5. Have your student solve a few additional examples.

⇒ $6.20 – $2.50   ($3.70)

⇒ $8.25 – $2.50   ($5.75)

⇒ $3.10 – $1.60   ($1.50)

(3) Subtract dollars and cents when the cents is almost $1

Activity

Use bills and coins. Ask your student to count out $5.60
Tell him that we are going to buy something that costs
95¢. How could we do that? We would use a $1 and get
5¢ change. Since 95¢ is almost a dollar, it is easy to know
how much change we will get.

So if we have a problems such as $5.60 – 95¢ we can find
the answer in the same way; subtract $1 and add back in
the 5¢ change. Write the equation.

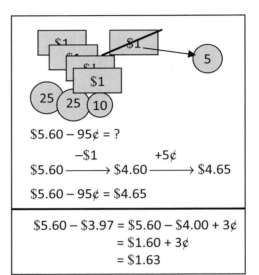

$5.60 – 95¢ = ?

$$-\$1 \qquad\qquad +5¢$$
$$\$5.60 \longrightarrow \$4.60 \longrightarrow \$4.65$$

$5.60 – 95¢ = $4.65

Write the problem **$5.60 – $3.97**. Ask your student what
change she would get by using a dollar to pay for
something that costs 97¢. To do this problem, we can
subtract $4, and then add back in 3¢.

$$\$5.60 - \$3.97 = \$5.60 - \$4.00 + 3¢$$
$$= \$1.60 + 3¢$$
$$= \$1.63$$

Discussion

Task 7, p. 56

7. (a) 3.60	3.65
	3.65
(b) 3.25	3.26
	3.26

Practice

Task 8, p. 56

Workbook

Exercise 14, p. 84 (Answers p. 73)

8. (a) $2.50	(b) $2.35
(c) $3.21	(d) $1.01

Reinforcement

Mental Math 42

(4) Solve word problems

Discussion

Tasks 9-10, p. 57

| 9. $1.55 |
| 10. $3.45 |

Make sure your student understands why the problems can be solved using subtraction.

Task 9: The total is what was paid for the doll. The two parts are the cost of the doll and the change. We need to find how much change she received.

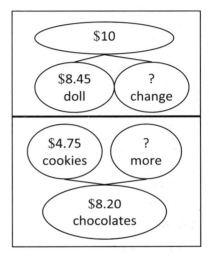

Task 10: The cookies cost some amount plus more to equal the cost of the chocolates.

You can also have your student do some of the word problems from the practices on pages 58-59 as part of this lesson.

Workbook

Exercise 15, pp. 85-86 (Answers p. 73)

The workbook problems involve both addition and subtraction. Your student needs to read the problem carefully, not simply subtract the two dollar amounts given.

Reinforcement

Use the store cards. Shuffle and place face down. Have your student turn over two at a time, and find the difference.

Extra Practice, Unit 9, Exercise 3, pp. 137-138

(5) Practice

Practice

   Practice B, p. 58

<table>
<tr><td>1. (a) $6</td><td>(b) $0.35</td></tr>
<tr><td>2. (a) $5.95</td><td>(b) $4.05</td></tr>
<tr><td>3. (a) $9.05</td><td>(b) $1.80</td></tr>
<tr><td>4. (a) $9.45</td><td>(b) $1.55</td></tr>
<tr><td>5. (a) $8.59</td><td>(b) $3.15</td></tr>
</table>

6. $1.40 + $7.85 = $9.25
   The total cost is **$9.25**.

7. $10 − $6.30 = $3.70
   The pair of slippers is **$3.70** cheaper.

8. $5 − $1.85 = $3.15
   He received **$3.15** change.

9. $4.25 + $1.95 = $6.20
   He has **$6.20**.

10. $5.65 + $1.70 = $7.35
    She has **$7.35** now.

   Practice C, p. 59

<table>
<tr><td>1. (a) $10</td><td>(b) $0.35</td></tr>
<tr><td>2. (a) $9.35</td><td>(b) $1.90</td></tr>
<tr><td>3. (a) $9.55</td><td>(b) $1.30</td></tr>
<tr><td>4. (a) $10.20</td><td>(b) $1.65</td></tr>
<tr><td>5. (a) $9.75</td><td>(a) $0.06</td></tr>
</table>

6. $6.80 + $2.40 = $9.20
   He had **$9.20** at first.

7. $1.95 + $1.60 = $3.55
   She spent **$3.55**.

8. $6.45 − $3.95 = $2.50
   He saved **$2.50** in the second week.

9. $8.05 − $1.90 = $6.15
   The best cost **$6.15**.

10. $9.20 − $2.80 = $6.40
    She had **$6.40** left.

Reinforcement

   Mental Math 43

Test

   Tests, Unit 9, Chapter 3, A and B, pp. 55-58

Workbook

Exercise 11, p. 81

1. (a) 1.85
   (b) 4.45
   (c) 3.05
   (d) 1.25
   (e) 2.35
   (f) 5.05
   (g) 6.00
   (h) 9.15
   (i) 3.20
   (j) 4.30
   (k) 2.45
   (l) 5.25

Exercise 12, p. 82

1. (a) 4.80   4.30
          4.30
   (b) 1.75   1.40
          1.40
   (c) 2.90   2.25
          2.25

2. (a) $3.60
   (b) $3.25
   (c) $2.15
   (d) $4.35
   (e) $2.45
   (f) $4.25

Exercise 13, p. 83

1. A $1.65   D $2.55   F $4.75
   G $2.30   L $2.85   N $4.60
   O $2.60   R $0.45   Y $1.55
   DRAGONFLY

Exercise 14, p. 84

1. (a) $3.31
   (b) $2.46
   (c) $2.26
   (d) $3.01

2. (a) $1.25
   (b) $2.60
   (c) $0.15
   (d) $2.30

Exercise 15, *p. 85*

1. $1 − 55¢ = 45¢
   He got **45¢** change.

2. $8 − $5.35 = $2.65
   He had **$2.65** left.

3. $5.90 - $3.85 = $2.05
   The doll is **$2.05** cheaper.

4. Cost of stamps = $2 + 75¢ + 20¢ = $2.95
   $2.95 + $6.30 = $9.25
   She had **$9.25** at first.

5. $2.60 + $0.95 = $3.55
   His brother spent **$3.55**.

6. $10 − $1.95 = $8.05
   Jose saved **$8.05**.

Review 9

Review

Review 9, pp. 60-61

You can do this review all at once, or you can select problems each day for more continuous review as you continue in the textbook.

For problem 7 your student might just look at the tens and give the answer as 4 tens. 34 tens is more accurate. To reduce ambiguity change the problem to read "How many tens in all are there in 349?" to give an answer of 34 tens, or change to problem to ask for hundreds and ones as well.

Your student may try to solve problem 10 as a division problem, and may find no amount left over, since 10 is divisible by 2. This one might be more easily solved with repeated subtraction, since there are only 4 presents.

Workbook

Review 10, pp. 87-91 (Answers p. 75)

Test

Tests, Units 1-9, Cumulative, A and B, pp. 59-65

---

1. (a) 833   (b) 300   (c) 479

2. (a) 505   (b) 101   (c) 127

3. (a) 24   (b) 50   (c) 24

4. (a) 4   (b) 9   (c) 9

5. 18 in. + 31 in. + 25 in. = **74 in.**
   The total length around is 74 in.

6. Centimeters

7. 34 tens

8. 5 lb: $45
   1 lb: $45 ÷ 5 = $9
   1 lb of shrimp costs **$9**.

9. 1 yd: $3
   9 yd: $3 x 9 = $27
   She paid **$27**.

10. 10 − 2 − 2 − 2 − 2 = 2
    2 ft of ribbon will be left over.

11. 132 cm − 119 cm = 13 cm
    John is **13 cm** taller.

12. $800 − $398 = $402
    The television cost **$402**.

13. 1 week: $3
    5 weeks: $3 x 5 = $15
    He saves **$15** in 5 weeks.

14. 450 m + 365 m = 815 m
    She walked **815 m**.

15. $5 − $2.45 = $2.55
    He received **$2.55** in change.

Workbook

Review 10, pp. 87-91

1. (a) 451
   (b) 960

2. (a) 999
   (b) 700
   (c) 908

3. (a) m
   (b) cm
   (c) cm
   (d) m

4. (a) kg
   (b) g
   (c) g
   (d) kg

5. (a) oz
   (b) in.
   (c) ft
   (d) lb

6. (a) 4 lb > 4 oz
   (b) 1 ft = 12 in.
   (c) 1 yd < 4 ft
   (d) 3 in. > 3 cm

7. (a) 1000   (b) 264
   (c) $5.90   (d) $2.80
   (e) $5.54   (f) $7.05

8. (a) 6; 9; 18; 24
   (b) 20; 32; 36; 40
   (c) 15; 20; 30; 35
   (d) 30; 40; 50; 70; 90

9. 1 box: 3 pens
   8 boxes: 3 x 8 = 24
   She bought **24** pens.

10. How many 5's in 23?
    23 ÷ 5 = 4 with 3 left over.
    (a) He cut 4 pieces.
    (b) The piece left over was **3 m** long.

11. $10 − $5.35 = $4.65
    He received **$4.65** in change.

12. 920 g − 135 g = 785 g
    The papaya weighs **785 g**.

13. 1 day: 6 kg
    4 days: 6 kg x 4 = 24 kg
    24 kg of meat are needed to feed the tigers for 4 days.

14. $9.50 − $1.60 = $7.90
    The pen costs **$7.90**.

Unit 10 – Fractions

Chapter 1 – Halves and Quarters

Objectives

- Recognize and name halves and quarters.
- Equate quarters to fourths.
- Read and write one half and one fourth as fractions.

Vocabulary

- One half
- Halves
- One fourth
- One quarter
- Fourths
- Quarters
- Fraction

Notes

Students were introduced to halves and quarters in *Primary Mathematics* 1B. This is reviewed in this chapter and the fractional notation $\frac{1}{2}$ and $\frac{1}{4}$ are introduced.

$\frac{1}{2}$ of a whole means 1 out of 2 equal parts, and $\frac{1}{4}$ of a whole means 1 out of 4 equal parts. Two halves or four fourths make a whole. Half of one whole is not necessarily the same as half of a different whole. If we want to compare $\frac{1}{2}$ and $\frac{1}{4}$, they must both be fractions of the same whole.

A "quarter" used in this context means the same as a fourth. Your student has recently heard of the term quarter in context of money. You may want to point out that since there are four quarters in a dollar, a quarter is a fourth of a dollar. In this and the next chapter the whole is one object, such as a pizza or a shape. Fractions of a set, where the whole is more than one object, will be introduced in chapter 3.

Material

- Paper strips, 4 the same width and length
- Sheets of paper
- Paper squares (cut out of a sheet of paper by folding a corner up and cutting off a strip)

(1) Identify halves and fourths

Activity

Ask your student what he would do if he wanted to share four cookies equally between himself and a friend? How would he share two cookies? What about one cookie? He would have to divide it into two equal pieces and give himself and his friend each a **half**. Ask him whether, if he gave his friend a little bit off of the cookie, he would be giving a half. No, both pieces must be the same size.

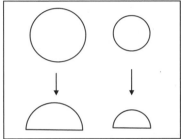

Draw a large circle and a small circle. You could tell your student these are two cookies, or two pizzas. Ask him to divide each one into half. Then, draw each half. Each drawing is a half, but each drawing is not the same. Ask your student how they can both be half, when they are not the same. They are both half of different wholes. When we talk about half *of something*, we always include the *of something*.

Give your student a strip of paper and ask her to cut it into two equal pieces. After she has made the cut, compare the two pieces by laying them on top of each other to see if they are the same. If they are not, then each piece is not a half. If necessary, show her how she can fold a strip in half to get equal pieces. Each piece is a half of the whole strip of paper. The two pieces have to be the same to each be a half.

Then, ask your student to cut another strip the same size as the first strip into four equal pieces. Again, compare the 4 pieces by laying them on top of each other. If he did not fold the strip of paper in half and half again to get four equal pieces, show him how with another strip. Tell him that each piece is a **fourth** of the whole. Each piece has to be equal for one piece to be a fourth.

Compare the strip that is half of a whole to the one that is a fourth of a whole. Ask your student which one is larger. If we want to compare a fourth to a half, they must be from the same whole. Otherwise the comparison has no meaning, since a fourth of a large strip could be larger than half of a small strip. So if we are asked whether a fourth is larger or smaller than a half, and are not told the whole, then we assume the whole is the same for both.

Discussion

Concept pages 62-63

Tasks 1-2, p. 63

1. (a) B, D   (b) P, Q

2. (a) 2   (b) 4

Activity

Tell your student that we say **one half** and write a 1 over a 2 to show 1 out of 2 equal parts of a whole: $\frac{1}{2}$. We say **one fourth** and write a 1 over a 4 to show 1 out of 4 equal parts of a whole: $\frac{1}{4}$. One half and one fourth are called **fractions** of a whole. A fourth is also sometimes called a **quarter**.

Fold one strip in half and ask your student to color one half. Fold another strip in fourths and ask your student to color one fourth.

Ask your student to write $\frac{1}{2}$ and $\frac{1}{4}$ on the colored part.

Workbook

Exercise 1, pp. 92-93 (Answers p. 79)

Reinforcement

Extra Practice, Unit 10, Exercise 1, pp. 143-144

Test

Tests, Unit 10, Chapter 1, A and B, pp. 67-72

Enrichment

Give your student some pieces of paper in the shapes of rectangles or squares and have her experiment with different ways to divide them into two or four equal parts. If she wonders if the halves are the same when they come from the same shape but look differently, you can show her they are the same by cutting up the halves and rearranging the pieces; the areas are the same.

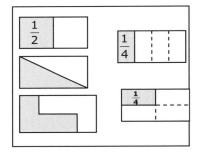

Workbook

Exercise 1, pp. 92-93

1. First and second one of first row
   First and third one of second row.

2. First one of first row;
   First, second and fourth one of second row.

3. Check work.

4. Check work.

5. (a) Check work.
   (b) >

Chapter 2 – Writing Fractions

Objectives

♦ Understand fractional notation.
♦ Identify fractions of a whole.
♦ Read and write fractions of a whole.
♦ Compare and order unit fractions.
♦ Make a whole with two or more fractions with the same denominator.

Vocabulary

♦ Third   ♦ Seventh   ♦ Tenth
♦ Fifth   ♦ Eighth   ♦ Eleventh
♦ Sixth   ♦ Ninth   ♦ Twelfth

Notes

In this chapter your student will learn how to represent the number of parts out of a total number of equal parts using fractions. $\frac{1}{4}$ represents 1 out of 4 equal parts of the whole. $\frac{3}{4}$ represents 3 out of 4 equal parts of the whole. The top number (numerator) counts the number of parts. The bottom number (denominator) tells us the number of parts the whole has been divided into.

Natural numbers count objects. The object being counted, such as apples or centimeters, is the denomination of the number. Fractions count *parts* of objects. The part from which the fraction is taken is called the *whole*. The whole for a fraction is like the denomination of a number. When we take three fourths of an apple, the whole is the apple, and the parts being counted are fourths. When we take three fourths of 12, the whole is 12, and the parts being counted are each a fourth of 12.

Fractions with a 1 in the numerator are called unit fractions. Your student will learn that the more parts the whole is divided into, the smaller the part. So $\frac{1}{6}$ is smaller than $\frac{1}{4}$.

Make sure your student understands that when we compare fractions, we are comparing the fractions of the *same* whole. One sixth of a square is smaller than one half of the same square, but is not necessarily smaller than one half of another square.

At this level, students do not learn the terms *numerator* and *denominator*.

The lessons in this chapter are fairly short and easy, so you may want to combine several lessons.

Material

♦ Fraction strips and circles Appendix p. a27-a30, 3 copies of each)
♦ Multilink cubes
♦ Index cards (for fraction cards, see lessons)

(1) Read and write fractions of a whole

Discussion

Concept page 64

For the first shape, ask your student how many parts it has been divided into. (Three) Tell her that it is divided into **thirds**. Read the fraction, **one third.** Repeat with the other two shapes as you go through this page. Use **fourth** rather than quarters in general, but tell her that $\frac{1}{4}$ can be read as **one fourth** or **one quarter**, and $\frac{3}{4}$ as **three fourths** or **three quarters**.

Tasks 1-2, p. 65

Make sure your student can read the fractions correctly.

Activity

Use a copy of the unlabeled fraction bars in the appendix. You can cut them apart in strips or leave them together. Ask your student to count the parts for each strip, color the first part, and write the fraction on the first part. Help your student read each fraction, one half through one twelfth.

Use another copy of the fraction bars or a copy of the fraction circles. Select a strip or a circle and ask your student to color a specific number of parts, e.g. 3 out of 7 parts, and then write the fraction that is colored.

Workbook

Exercise 2, pp. 94-95 (Answers p. 85)

1. (a)	1 out of 5
(b)	4 out of 5
2. (a)	$\frac{1}{6}$
(b)	3 out of 8
	$\frac{3}{8}$

$\frac{1}{2}$	one **half**
$\frac{1}{3}$	one **third**
$\frac{1}{4}$	one **fourth**
$\frac{1}{5}$	one **fifth**
$\frac{1}{6}$	one **sixth**
$\frac{1}{7}$	one **seventh**
$\frac{1}{8}$	one **eighth**
$\frac{1}{9}$	one **ninth**
$\frac{1}{10}$	one **tenth**
$\frac{1}{11}$	one **eleventh**
$\frac{1}{12}$	one **twelfth**

(2) Identify and write fractions of a whole

Discussion

Tasks 3-4, p. 66

Have your student write the fractions for task 4. Be sure he understands that the top number is the number of colored parts and the bottom number is the total number of parts of the whole.

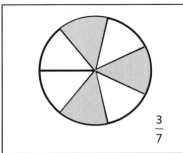

3. (a) $\frac{3}{7}$

  (b) $\frac{5}{6}$

4. (a) $\frac{4}{9}$ (b) $\frac{4}{6}$

  (c) $\frac{5}{12}$ (d) $\frac{7}{10}$

Activity

Use the fraction bars or circles. Color in some parts, not necessarily consecutive, and have your student write the fraction that is colored.

Use multilink cubes of two different colors, such as red and yellow. Put up to 12 of them together to form a shape. Ask your student what fraction of the shape is red (or yellow).

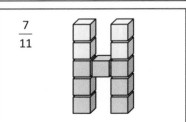

$\frac{3}{7}$

Write a fraction and have your student create a shape using two colors of multilink cubes where one of the colors is that fraction of the whole.

$\frac{7}{11}$

Workbook

Exercise 3, pp. 96-99 (Answers p. 85)

Enrichment

Get your student to estimate with fractions:

Draw some bars the same size as the whole in the fraction bars in the appendix, or commercial fraction bars you might have. Color a portion of a bar and ask your student about how much is shaded. Then have him check his estimation with the fraction bars. You can give him three choices and have him choose the best estimate.

About $\frac{4}{5}$ of the bar is shaded.

(3) Compare and order unit fractions

Activity

Draw two circles the same size. Tell your student that 6 people want to share a pizza. Divide one of the circles into 6 pieces. Then tell him that 4 people want to share a pizza of the same size. Divide the other circle into 4 pieces. Ask who would get a larger piece, someone in the first group or someone in the second group? Why? Since the first pizza has to be cut up into more pieces, each piece is smaller.

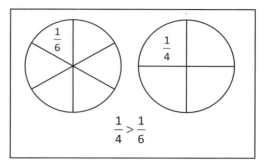

Ask your student whether the one sixth piece would be smaller than the one fourth piece if the pizza it is cut from is larger. It might not be; we can only compare fractions if the whole for each is the same size. So when we say that one fourth is greater than one sixth, we have to assume that the whole for both fractions is the same shape and size.

Color one of each of the pieces of the fraction bars copied from appendix p. a28 and cut the strips apart. Mix them up and have your student put the colored parts in order. Then ask her if she notices anything about the fractions. Point out, if necessary, that smaller fractions have a larger number for the total number of parts, i.e. the bottom number is larger. See if she can tell you why. If the whole is cut up into more parts, each part has to be smaller. Give her several of the bars and have her put them in order.

Write down some unit fractions on index cards. Tell your student that they all represent fractions of the same whole. Ask him to put them in order.

Practice

Tasks 5-6, p. 67

Activity

Use a copy of the fraction bars or circles. They do not need to be cut apart.

Name two fractions, not unit fractions, such as $\frac{3}{4}$ and $\frac{4}{5}$. Have your

student select the appropriate strips or circles, color in the fraction, compare them, and tell you which is smaller or larger. Repeat with other pairs of fractions.

5. $\frac{1}{4}$

6. $\frac{1}{8}$, $\frac{1}{5}$, $\frac{1}{2}$

Workbook

Exercise 4, pp. 100-102 (Answers p. 85)

Game

Material: 4 sets of unit fractions, $\frac{1}{2}$, $\frac{1}{3}$, $\frac{1}{4}$, $\frac{1}{5}$, $\frac{1}{6}$, $\frac{1}{7}$, $\frac{1}{8}$, $\frac{1}{9}$, $\frac{1}{10}$, $\frac{1}{11}$, $\frac{1}{12}$

Procedure: Shuffle cards and place them face down in the middle. For each round, players take turns drawing a card and turning it over. The one with the largest fraction gets all the cards. If it is a tie, the tied players draw another card and compare. The player with the most cards at the end, when all the cards are turned over or there are not enough for each player to turn one over wins.

(4) Find fractions that make a whole

Activity

Refer to the pictures in tasks 3-4, p. 66. For the first shape, ask your student for the fraction of the shape that is colored and write it down. Then ask her for the fraction that is not colored and write it down. Ask her what both fractions together make. The uncolored part plus the colored part make a whole, so the fractions representing the colored part and uncolored part together make a whole.

$\dfrac{3}{7}$ and $\dfrac{4}{7}$ make 1 whole.

$\dfrac{7}{7}$ = 1 whole

Ask your student what the top number represents for each of the two fractions. They represent the number of parts. Then ask him to add the top numbers together. The total is the same as bottom number of each fraction. Ask him what the fraction $\dfrac{7}{7}$ means. All 7 parts out of 7 is the same as the whole.

For the rest of tasks 3 and 4, have your student find the fraction that is colored and the fraction that is not colored. Write each one down and let her determine the total of the top numbers to see if it matches the bottom numbers.

Discussion

Task 7, p. 67

7. (a) $\dfrac{2}{5}$

(b) $\dfrac{6}{7}$

(c) $\dfrac{7}{9}$

Activity

Write some other fractions down and ask your student to find the fraction that makes 1 whole with it without the use of a drawing.

Workbook

Exercise 5, pp. 103-104 (Answers p. 85)

Reinforcement

Extra Practice, Unit 10, Exercise 2, pp. 145-150

Test

Tests, Unit 10, Chapter 2, A and B, pp. 73-80

Game

Material: A set of fraction cards with the following fractions: $\dfrac{1}{2}, \dfrac{1}{2}, \dfrac{1}{3}, \dfrac{2}{3}, \dfrac{1}{4}, \dfrac{2}{4}, \dfrac{3}{4}, \dfrac{1}{5}, \dfrac{2}{5}, \dfrac{3}{5}, \dfrac{4}{5},$

$\dfrac{1}{6}, \dfrac{2}{6}, \dfrac{3}{6}, \dfrac{3}{6}, \dfrac{1}{7}, \dfrac{2}{7}, \dfrac{3}{7}, \dfrac{4}{7}, \dfrac{5}{7}, \dfrac{6}{7}, \dfrac{7}{7}, \dfrac{1}{8}, \dfrac{2}{8}, \dfrac{3}{8}, \dfrac{4}{8}, \dfrac{4}{8}, \dfrac{5}{8}, \dfrac{6}{8}, \dfrac{7}{8}, \dfrac{8}{8}, \dfrac{1}{9}, \dfrac{2}{9}, \dfrac{3}{9}, \dfrac{4}{9}, \dfrac{5}{9}, \dfrac{6}{9}, \dfrac{7}{9}, \dfrac{8}{9}, \dfrac{9}{9}$

Procedure: Shuffle cards and place face down in the middle. Turn over the top one and put it face up in the middle. Players take turns turning over a card. If there is a card in the middle that makes a whole with the card they turned over, the keep both cards. If there is not, they place their card face up in the middle. Play continues until all cards are paired.

Workbook

Exercise 2, pp. 94-95

1. (a) $\frac{2}{3}$

   (b) $\frac{5}{8}$

   (c) $\frac{7}{10}$

   (d) $\frac{3}{4}$

2. (a) 1 out of 6
   (b) 2 out of 5
   (c) 1 out of 3
   (d) 3 out of 4
   (e) 5 out of 8

Exercise 3, pp. 96-97

1. Check answer.

2. The lines form a cross-hatched picture with the fish in the middle of each diamond shape.

3. A $\frac{3}{4}$ C $\frac{2}{3}$ F $\frac{3}{5}$

   I $\frac{5}{6}$ O $\frac{1}{6}$ N $\frac{2}{5}$

   R $\frac{1}{2}$ S $\frac{5}{12}$ T $\frac{3}{8}$

   FRACTIONS

4. Check the fractions colored.

Exercise 4, pp. 100-102

1. Check work.

2. (a) >   (b) <
   (c) <   (d) >
   (e) <   (f) >

3. (a) $\frac{1}{3}$ (b) $\frac{1}{4}$

   (c) $\frac{1}{9}$ (d) $\frac{1}{2}$

   (e) $\frac{1}{8}$ (f) $\frac{1}{4}$

4. (a) $\frac{1}{6}$ (b) $\frac{1}{10}$

   (c) $\frac{1}{5}$ (d) $\frac{1}{12}$

   (e) $\frac{1}{10}$ (f) $\frac{1}{9}$

5. (a) $\frac{1}{2}$ (b) $\frac{1}{5}$

   (c) $\frac{1}{4}$ (d) $\frac{1}{5}$

6. (a) $\frac{1}{7}$ (b) $\frac{1}{12}$

   (c) $\frac{1}{4}$ (d) $\frac{1}{9}$

7. $\frac{1}{10}$ $\frac{1}{8}$ $\frac{1}{4}$ $\frac{1}{2}$

8. $\frac{1}{3}$ $\frac{1}{5}$ $\frac{1}{9}$ $\frac{1}{12}$

Exercise 5, pp. 103-104

1. (a) $\frac{2}{3}$

   (b) $\frac{6}{8}$

   (c) $\frac{3}{5}$

   (d) $\frac{5}{6}$

2.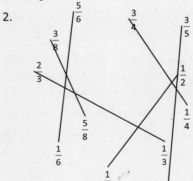

Chapter 3 – Fraction of a Set

Objectives

- Name parts of sets as fractions of a whole when each part has 1 item.
- Find the fraction of a set when each part has 1 item.
- Name parts of sets as fractions of a whole when each part has more than 1 item.
- Find the fraction of a set when each part has more than 1 item.
- Find how many items are in a given fraction of a set.

Notes

In the previous two chapters the whole was a geometric shape or a bar. In this chapter your student will learn about fractions of a set, where the whole is a set of objects. The objects are divided into equal parts. Again, the numerator counts the number of parts, and the denominator is the total number of equal parts the whole has been divided into.

For example, a set of 8 objects is divided into 4 equal groups of 2. One of those groups of 2 is $\frac{1}{4}$ of the whole. 3 of those groups is $\frac{3}{4}$ of the whole. As with fractions of geometric shapes, a fourth of one whole is not necessarily equal to a fourth of a different whole. $\frac{1}{4}$ of 8 is not the same as $\frac{1}{4}$ of 12. If we were to compare fractions

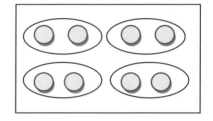

of a set, we could only compare fractions with the same value for a whole. So $\frac{1}{2}$ of 8 is greater than $\frac{1}{4}$ of 8, but $\frac{1}{2}$ of 2 is less than $\frac{1}{4}$ of 8. Fractions have no meaning without the whole.

To find the value of $\frac{3}{4}$ of 8 we can first find the value of $\frac{1}{4}$ of 8. $\frac{3}{4}$ is then 3 of those parts. The denominator tells us how many equal parts there are and the numerator tells us how many of those parts to count.

Your student may realize that there are fractions that have the same value; e.g. two eighths is the same as one fourth. You can tell him they are; two eighths of 8 is 2, and one fourth of 8 is also 2. Equivalent fractions will be taught in *Primary Mathematics* 3.

Fractions express division. To find one fourth of 8, we divide 8 into 4 parts. You can use the language of division with your student and she will intuitively realize that we can divide to find the fraction of a set, but the equivalence of fractions and division will be formally taught in *Primary Mathematics* 4.

This is a concrete and pictorial introduction to fraction of a set. Do not require your student to find the fraction of a set without drawings or counters at this level.

Material

- Multilink cubes
- Counters
- Appendix p. a31

(1) Find the fraction of a set when each part has 1 item

Activity

Discuss ways in which fractions are used in daily life. For example, half the people in the family might be girls. Or about a fifth of the yard is covered with gravel.

Use multilink cubes. Use one block of one color, such as red, and 3 of another. Put them together. Ask your student what fraction of the shape is red. One fourth of the shape is red. The whole is the box shape formed from the 4 blocks.

Now, separate the blocks, and ask your student what fraction of the blocks is red. One fourth of the blocks is red. Tell your student that this time the whole is 4 blocks. Each block is an equal part. 1 out of 4 of the blocks are red, so we can say that one fourth of the blocks are red.

Show your student 5 objects, 2 of one kind and 3 of another, such as counters and blocks. Tell him that all 5 objects are the whole. Each object is a part. Then ask him what fraction of the whole are blocks. Two fifths are blocks. Ask him what fraction are counters. Three fifths are counters. Ask him if two fifths and three fifths make one whole. Yes, the blocks and the counters together make the whole of 5 objects.

Tell your student that a fraction of a whole can be a fraction of any kind of whole. The whole could be a pizza, or the whole could be a total of 4 blocks, or the whole could be 5 different objects.

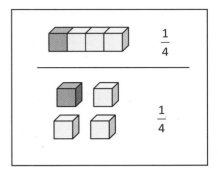

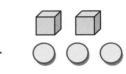

$\frac{2}{5}$ of the objects are blocks.

$\frac{3}{5}$ of the objects are counters

$\frac{2}{5}$ and $\frac{3}{5}$ make 1 whole

Discussion

Tasks 1-7, pp. 69-70

Tasks 1-3: Ask your student what fraction the other part is. In Task 1, three fourths of the blocks are purple. In Task 3, three fifths of the flowers are tulips.

Task 5: Point out that when we are told that there are 5 toys, $\frac{2}{5}$ of the toys are cars, so we know that 2 out of the 5 are cars. Ask how we can find how many are planes if we did not have a picture? We would subtract the number of cars from the number of toys.

Workbook

Exercise 6, pp. 105-107

Reinforcement

Extra Practice, Unit 10, Exercise 3, pp. 151-152

2. $\frac{1}{3}$

   $\frac{2}{3}$

3. $\frac{2}{5}$

4. $\frac{5}{8}$

   $\frac{5}{8}$

5. 3

6. 1

7. $\frac{3}{4}$

(2) Identify the fraction of a set

Activity

Give your student three counters of one color, such as green, and 9 of another color, such as blue. Ask him how many counters he has. Ask him to group them by 3, putting all the green counters in one group. Then ask him what fraction of the counters are green. He may say three twelfths. Tell him that is correct, when we take the whole to be 12, and each equal part to be 1 out of the 12.

Ask your student how many equal parts there are. There are 4 equal parts. Tell her that since each part is equal, we can use the number of parts as the whole. 1 part has green counters. So 1 part out of 4 parts is green. We can say that one fourth of the counters are green.

Ask your student what fraction of the counters are blue. Guide him to say that three fourths of the counters are blue.

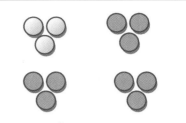

$\dfrac{1 \text{ group of green counters}}{4 \text{ groups of counters total}}$

$\dfrac{1}{4}$ of the objects are green.

$\dfrac{3}{4}$ of the counters are blue.

Discussion

Concept page 68

Tasks 8-14, pp. 71-72

Compare the answers for 11 and 12. Both are half, but the value for a half in each is different because the whole is different.

For task 13, ask, "If 2 dolls are one third of the total dolls, how can we find the total number of toys without counting or seeing a picture?" We know there are 3 equal groups, and that one out three has 2 toys. So we can multiply to find the total number: 2 x 3 = 6.

Workbook

Exercise 7, p. 108

Reinforcement

Give your student some counters, of which a fraction is a specific color. Ask him to make equal groups and then tell you the fraction that is that color, using the number of groups of the whole. For example, give him 20 counters of which 5 are yellow and the rest are any other colors and have him tell you what fraction are yellow.

Enrichment

Give your student 12 counters and ask her to find $\dfrac{1}{4}$ of the whole. Since the bottom number is 4, she needs to make 4 equal groups. Then, $\dfrac{1}{4}$ of 12 is the number that is in each group, 3. Then ask her to find $\dfrac{3}{4}$ of 12. This is the number in 3 of the 4 equal groups. Repeat with other examples, asking her to find the fraction of a whole using counters. Only use examples where the whole can be evenly divided by the number in the denominator.

8. $\dfrac{1}{3}$

9. $\dfrac{1}{4}$ are pink

$\dfrac{3}{4}$ are blue

10. $\dfrac{2}{5}$ are yellow

11. $\dfrac{1}{2}$ are apples

5 apples

12. $\dfrac{1}{2}$

8

13. 2

14. 3

(3) Find what fraction of a set a number is

Activity

Give your student 20 counters all of the same color. Ask him, "4 is what fraction of 20?" Tell him you want a fraction where the top number is 1; you want 4 as one part of the whole. He needs to group the counters by 4 and count the number of groups. There are 5 groups, each with 4 counters. 1 group out of 5 groups is one fifth, so 4 is one fifth of 20.

Do the enrichment below, if you want, and then repeat the above activity with other examples as needed.

Enrichment

Ask your student, "8 is what fraction of 20?" We can't make equal groups of 8, but from the groups of 4, we can see that 8 is two fifths of 20.

Ask your student, "How many counters are there in three fifths of 20?"

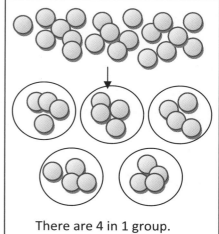

There are 4 in 1 group.
There are five groups.

4 is $\frac{1}{5}$ of 20

Discussion

Task 15-17, p. 73

If needed, use counters to represent the oranges in task 17.

Workbook

Exercise 8, p. 109

Reinforcement

Extra Practice, Unit 10, Exercise 3, pp. 151-152

Test

Tests, Unit 10, Chapter 3, A and B, pp. 81-88

15.	$\frac{1}{5}$
16.	$\frac{1}{4}$
17.	$\frac{1}{5}$
	$\frac{4}{5}$

Enrichment

Appendix p. a31. Have your student make groups in order to answer the questions. The fractions must have 1 in the top number.

Answers:

What fraction of the shapes are circles? $\frac{1}{4}$

What fraction of the shapes are striped? $\frac{1}{3}$

What fraction of the shapes are dotted? $\frac{1}{6}$

What fraction of the shapes are have 4 sides? $\frac{1}{2}$

What fraction of the shapes are black? $\frac{1}{12}$

Workbook

Exercise 6, pp. 105-107

1. 5 equal parts
   1 part kittens
   4 parts puppies

   $\frac{4}{5}$ of the pets are puppies.

2. (a) $\frac{3}{4}$

   (b) $\frac{3}{8}$

   (c) $\frac{2}{9}$

3. (Which specific shells are colored can vary.)

   (a)

   (b)

   (c)

   (d)

4. $\frac{2}{5}$

5. (a) $\frac{5}{8}$

   (b) $\frac{3}{8}$

6. 1 out of 5 were wrong, so 4 out of 5 were right.

   He got $\frac{4}{5}$ of the problems right.

7. She gave 2 out of 6 to her friend, so she had 4 out of 6 left.
   She had **4** sandwiches left.

8. 5 out of 11 were girls, so 6 out of 11 were boys.
   There were **6** boys.

Exercise 7, p. 108

1. 1 part apples $= \frac{1}{5}$
   5 equal parts

2. The set of circles should be circled.

3. (a) $\frac{3}{4}$

   (b) $\frac{1}{3}$

   (c) $\frac{2}{3}$

Exercise 8, p. 109

1. $\frac{1}{3}$

2. $\frac{1}{4}$

3. $\frac{1}{5}$

Review 10

Review

Review 10, pp. 74-75

Allow your student to use counters with problem 10.

Workbook

Review 11, pp. 110-114

In problem 11, the coins are not proportional in size.

Test

Tests, Units 1-10, Cumulative, A and B, pp. 89-100

Enrichment

Give your student a piece of string. Ask him how he can cut it into 4 equal pieces so that each piece is one fourth of the whole string. He can fold it in half, cut at the fold, and then fold each piece in half again and cut at the fold. Ask him how many places the string had to be cut in order to cut it into four pieces. It was cut in 3 places. Ask him how many places the string would be cut to cut it into five equal pieces? (4.) 6 equal pieces? (5.) Twenty equal pieces? (19.)

Draw a picture of a circular cake, or use a real cake (don't show him the ones drawn here). Ask your student how many cuts he would use to cut it into two equal pieces. (1.) Four equal pieces? (2.) Eight equal pieces? (4.) 16 equal pieces? (8.)

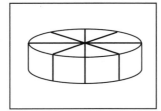

Challenge your student to figure out a way of cutting it into eight equal pieces with only 3 cuts. If she needs help, remind her of the different ways of dividing the paper into halves. What is another way she can cut the cake in half?

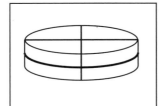

1. (a) 562   (b) 397   (c) 448

2. (a) 269   (b) 231   (c) 19

3. (a) 35   (b) 90   (c) 32

4. (a) 9   (b) 5   (c) 6

5. (a) 208   (199 + 9)
   (b) 530   (480 + 50)
   (c) 194   (202 – 8)
   (d) 283   (313 – 30)

6. $\frac{6}{9}$

7. $\frac{9}{12}$

8. $\frac{6}{6}$

9. $\frac{2}{3}$

10. (a) 20 ÷ 5 = 4
    There are **4** children in group A.
    (b) $\frac{1}{5}$

11. 38 ft ÷ 5 = 7 ft with 3 ft left over
    3 ft of ribbon are left over.

12. 520 – 485 = 35
    There are **35** more boys than girls.

13. 735 + 65 = 800
    Dan collected **800** stamps.

14. How many 3's in 27?
    27 ÷ 3 = 9
    He had **9** bags.

15. 5 m x 6 = 30 m
    The total length is **30 m**.

16. 1 pack: 4 batteries
    6 packs: 4 x 6 = 24 batteries
    There were **24** batteries altogether.

Workbook

Review 11, pp. 110-114

1. $\dfrac{5}{8}$

2. Check work.

3. $\dfrac{3}{4}$

4. $\dfrac{3}{8}$

5. (a) Cake B
   (c) Cake A

6. 2 pairs out of 3 pairs are brown.
   $\dfrac{2}{3}$ of her shoes are brown.

7. (a) 7   7
   (b) 8   8
   (c) 9   9
   (d) 8   8

8. (a) 405   413
   (b) 596   600
   (c) 402   398
   (d) 866   858

9. (a) 506   (b) 707
   (c) 802   (d) 394

10. (a) $0.40   (b) $0.80
    (c) $0.55   (d) $2.15

11. Bag A: $7.10   Bag B: $4.50
    $7.10 − $4.50 = $2.60
    There is $**2.60** more in bag A than in bag B.

12. 4 girls spent: $36
    1 girl spent: $36 ÷ 4 = $9
    Each of them paid **$9**.

13. 1 T-shirt: $4
    6 T-shirts: $4 x 6 = $24
    He paid **$24**.

14. 45 kg in groups of 5 kg
    45 kg ÷ 5 kg = 9
    There were **9** bags of sugar.

15. 28 − 19 = 9
    Mrs. Lambert bought **9** more mangoes.

16. 1 week: 7 words
    5 weeks: 7 x 5 = 35 words.
    He learns to spell **35** words in 5 weeks.

17. 26 + 34 = 60
    There were **60** books in the library at first.

Unit 11 – Time

Chapter 1 – Telling Time After the Hour

Objectives

♦ Develop a feel for a time duration of 1 hour.
♦ Develop a feel for a time duration of 1 minute.
♦ Tell time as minutes after the hour.
♦ Understand quarter hours and half hours as fraction of an hour.

Vocabulary

♦ O'clock
♦ Half past
♦ Quarter past
♦ Minute hand
♦ Hour hand

Notes

Students learned to tell time to the hour and half-hour and to read and say the time as "4 o'clock" or "half past 4" in *Primary Mathematics* 1B. In this chapter they will learn to tell time to the five-minute interval, to say the time as number of minutes after or past the hour, and to read and write time using the hour:minutes notation (e.g. 7:05). You can extend the concepts in this section to telling time to the minute.

Although the times in this chapter are to the 5-minute mark, your student will be able to tell time to the minute. If she is familiar with the number of minutes for each 5-minute mark on the clock, it will be easy to count up another several minutes if exact time is needed.

Usually, exact time is not needed, and many things start or end at times where the minutes are multiples of 5. When we look at an analog clock, we don't usually have to read it to the exact minute. The lessons will include teaching your student to read the time to the nearest 5 minutes.

Your student may be more familiar with a digital clock. An analog clock, or "face" clock, will give him a visual picture of time and the fraction of an hour that has passed. An analog clock is also useful in determining time intervals, particularly those that change from one hour to the next. Use a face clock with geared hands so that one hand moves in relation to the other. Because of the prevalence of digital clocks, time is usually told by simply reading the hours and minutes on the digital clock, rather than saying "half past" or "quarter past." Your student will still gain a better understanding of time, and ability to calculate elapsed time, by learning to tell time on an analog clock.

Material

♦ Analog clock with geared hands
♦ Large demonstration clock, if available
♦ Stopwatch

(1) Tell time after the hour

Activity

Ask your student how we measure time during the day. (We use a clock; we measure in hours and minutes.) See if he has a feel for how long an hour is. You can ask him for some activities that take about an hour. Then see if he has a feel for how long a minute is. You can use a stopwatch and time an activity for one minute, such as hopping in place. You can also discuss how time seems to change; an hour doing something fun seems to pass much faster than an hour doing something boring. That is because when we are bored we pay more attention to time passing; it does not really pass more slowly.

Show your student an analog clock. Tell her that the pointers on the clock are called "hands." Remind her that as an hour passes the long hand moves all the way around the clock. This hand measures the minutes. Set the time for 12:00 and move the minute hand all the way around. Ask her what happened to the short hand. It moves from the 12 to the 1. The short hand measures the hour at the same time that the long hand measures the minutes. Tell her we call the long hand the "minute hand" and the short hand the "hour hand."

Ask your student how many intervals there are between two numbers on the clock. If she counts on from one of the numbers to the next, there are 5 intervals marked. Tell her each interval is a minute. When the minute hand moves from one mark to the next, a minute has passed. Move the minute hand from number to number as your student counts by 5's. At 60, the minute hand has gone all the way around. There are 60 minutes in an hour.

Set the time on an hour, such as 6:00 and ask your student for the time. Write it down using the digital notation. Tell him that this is how we can write 6 o'clock. The colon (the two dots) separates the hours from the minutes. When the minute hand is straight up, no minutes have passed for the hour yet, so the minutes are 00. We read this time as "6 o'clock" or sometimes, when the other person knows we are talking about the time, simply as "six."

Move the minute hand to the 1. Ask your student how many minutes pass when the clock's minute hand moves from 12 to 1. Write the time as 6:05. Tell your student we read this as "six oh five" or "5 minutes past 6 o'clock" or "5 minutes after 6" or simply "5 past 6." Point out that as with money (e.g. the cents in $6.05) we need to have two digits for the minutes, so if there are no tens we put in 0 for the tens.

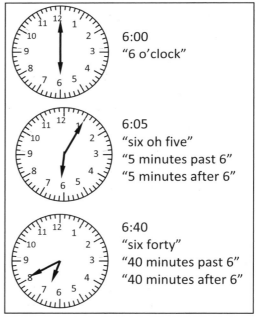

6:00
"6 o'clock"

6:05
"six oh five"
"5 minutes past 6"
"5 minutes after 6"

6:40
"six forty"
"40 minutes past 6"
"40 minutes after 6"

Continue around the clock for every five minutes, and have your student say the time. Point out that the hour hand is moving slowly around as well, so that by the time the minute hand is again pointing straight up, the hour had is pointing at the 7, and it is 7 o'clock.

Dictate a time as minutes past the hour and let your student set the corresponding time on the clock.

Discussion

Concept pages 76-77

Task 1, p. 77

1. (a) 5
(b) 30

Activity

Set some times on the clock and have your student read the time and write it in digital notation.

Set the time on a clock to 2:00 and then move the minute hand to 2:30. Tell your student that the minute hand has gone half-way around the circle. We can use "half past" if the minutes is 30. Have him look at the clock that "says" 2:30 on p. 77 of the text. If we put 60 items into two groups, how many would be in each group? 30 is in one group out of 2, so 30 is half of 60.

Now have your student look at the clock that says 9:15. Ask her what fraction of the circle the minute hand has gone around from the top on that clock. Guide her in seeing that it has gone one fourth, or one quarter of the way around. We can also say "quarter past" if the minutes is 15. Show 2:00 on the clock, move the minute hand to the 3, and ask her for the time (2:15, a quarter past 2). Then move the minute hand to the 9. Tell her three-quarters of an hour pass when the minute hand goes from straight up to point at the 9.

Tell your student that if we can remember the times when the minute hand points to the 3, the 6, or the 9, it is easy to tell the other times, because we can count up from those times. Set the time to 2:35 and ask her the time, getting her to count up from 2:30.

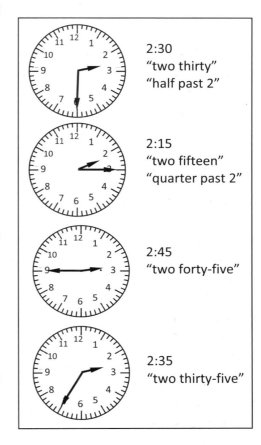

2:30
"two thirty"
"half past 2"

2:15
"two fifteen"
"quarter past 2"

2:45
"two forty-five"

2:35
"two thirty-five"

Set some times where the minute hand is pointing between the five-minute marks. Get your student to tell you which 5-minute mark it is closest to, and tell you the time to the closest 5-minutes. For example, set the time to 2:26 and get your student to say "It is about 2:25."

Workbook

Exercise 1, pp. 115-118 (Answer p. 96)

Reinforcement

Write some times, including times to the minute, in the digital format and have your student show the times on the clock, and tell you the time to the closest 5 minutes.

Extra Practice, Unit 11, Exercise 1, pp. 157-158

Test

Tests, Unit 11, Chapter 1, A and B, pp. 101-106

Workbook

Exercise 1, pp. 115-116

1. clockwise from top:
   0; 5; 10; 15; 20; 25; 30; 35; 40; 45; 50; 55

2. (a) 1:00   20
                  1:20
   (b) 4:00   15
                  4:15
   (c) 10:00   35
                  10:35

3.

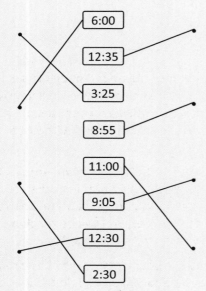

6:00

12:35

3:25

8:55

11:00

9:05

12:30

2:30

4. 7:30   8:15   1:30
   5:05   11:40   9:10
   1:20   8:50   12:45
   4:25   6:40   9:50

Chapter 2 – Telling Time Before the Hour

Objectives

♦ Tell time as minutes before the hour.
♦ Relate daily activities to the time.
♦ Understand the abbreviations a.m. and p.m.

Vocabulary

♦ Quarter of
♦ Midnight
♦ Noon
♦ a.m.
♦ p.m.

Notes

In this chapter your student will read and tell time as minutes before the hour. Though this is not used with digital clocks, learning how to tell time before the hour with an analog clock will help him understand the measurement of time and calculate elapsed time better. Using an analog clock gives him a visual understanding of how time is measured, which cannot be accomplished with just a digital clock.

In this chapter your student will relate events of the day to the time. In *Primary Mathematics* 1B, they learned that there are 24 hours in the day, measured in two groups of 12 hours. In this unit they will use the abbreviation a.m. (ante-meridiem) for before noon, and p.m. (post meridiem) for after noon. Since you will be discussing the hours of the day in this chapter, you will be teaching these terms now, even though they are not used until the next chapter with elapsed time.

You may want to include a discussion of telling time based on a 24-hour clock, in which the day runs from midnight to midnight and is divided into 24 hours numbered from 0 to 23. In the 24-hour notation, the day begins at midnight, 00:00, and the last minute of the day begins at 23:59. This system is the most commonly used time in the world today. In the U.S., 24-hour notation is referred to as military time. In spoken English, time on the hour is read as "o'clock" or as "hundred", e.g. 18:00 is "eighteen hundred" simply because it looks like hundreds in the decimal system, even though there are not a hundred minutes in an hour. 10:00 is "ten hundred" and not "one thousand" and 20:00 is "twenty hundred." In the U.S. military, time before 10:00 is read with "oh" for the hours, e.g. 03:00 is "oh three hundred." In common usage, the "oh" is not used. 00:00 is midnight, and 00:30 is often "midnight thirty" but could be "half past midnight."

Material

♦ Analog clock with geared hands
♦ Large demonstration clock, if available

(1) Tell time before the hour

Activity

Set the time on a demonstration clock or other analog clock with geared hands to a time that is after 30 minutes, such as 2:35. Have your student read the time. Tell her that when the time is after half-past the hour, we can also read the time as the number of minutes to the hour. Ask her how many minutes it is before three o'clock. She can either count by 5's backward around the clock from the 12, or forward from the 7. Tell her that we can read the time as "twenty-five minutes to three" or "twenty-five minutes before three" or simply as "twenty-five to three."

Set the time to a quarter before the hour, such as 2:45. Remind your student that we can divide the distance around the clock into fourths, or quarters. So this time can be read either as "fifteen minutes to 3" or "a quarter to three."

Set some times for five-minute intervals during the second half-hour and get your student to read the time as minutes to the hour. Then, set some times within the 5-minute marks and guide her in telling the time to the closest 5-minutes before the hour. You can also have him tell you the time to the minute.

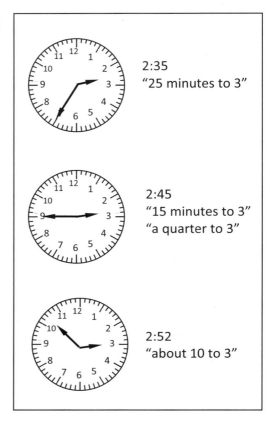

2:35
"25 minutes to 3"

2:45
"15 minutes to 3"
"a quarter to 3"

2:52
"about 10 to 3"

Discussion

Concept p. 78

Tasks 1-2, pp. 78-79

    For task 2, get your student to say the time in various ways, such as "six forty-five," "fifteen to seven," or "a quarter to seven" for 6:45. You may also want her to write the time in digital notation, or write it for her.

1. (a) 15		
(b) 5		
2. 6:00		6:05
6:20		6:45
7:15		7:50

Activity

Use a clock with geared hands. Show the time 12:00. Tell your student that this is 12 midnight, in the middle of the night. 12 midnight is considered to be the start of one full "day" which includes both night time and day time.

Go through various regular activities of the day in order, such as getting up, eating meals, etc. Have her show the time for each activity by moving only the minute hand until the correct time. She should note that the hours go from 12:00 midnight to 12:00 again at noon, which is the middle of the day. Tell her that 12:00 in the middle of the day is called "noon." After she shows her bed-time, continue moving the minute hand around to 12. Tell her this is now 12:00 midnight again. A new day starts.

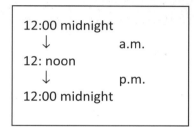

12:00 midnight
↓   a.m.
12: noon
↓   p.m.
12:00 midnight

Explain to your student that we use a.m. to stand for the times between 12:00 midnight and 12:00 noon and we use p.m. to stand for the times between 12:00 noon and 12:00 midnight. Write the abbreviations.

Pick an activity that could be done in the morning or in the evening, for example at 8:00. Ask your student what parts of the day each of the two 8:00's are in (morning and evening). Tell your student that in order to be able to tell someone else which part of the day it is, we need to say 8:00 a.m., if it is in the morning, or 8:00 p.m., if it is in the evening. We could also say, "Eight in the morning" or "eight in the evening." Sometimes it is obvious, such as, an invitation to come to a party starting at 2:00. This would likely not be 2:00 a.m. Ask your student why not.

Ask your student how many hours there are from 12:00 midnight to 12:00 noon and from 12:00 noon to 12:00 midnight. For each there are 12 hours. Ask him how many hours there are in the day. There are 24 hours in the day.

Workbook

Exercise 2, pp. 119-121 (Answers p. 100)

Reinforcement

Talk about time during the day, drawing your student's attention to the time on a real analog clock, and discuss whether it is a.m. or p.m.

Extra Practice, Unit 11, Exercise 2, pp. 159-160

Test

Tests, Unit 11, Chapter 2, A and B, pp. 107-113

Workbook

Exercise 2, pp. 119-121

1. (a) **5** minutes after **3**
       5 minutes past **3**
   (b) **25** minutes after **12**
       25 minutes past **12**
   (c) **20** minutes before **3**
       20 minutes to **3**
   (d) **10** minutes before **5**
       10 minutes before **5**

2. (a) **10** minutes past **6**
   (b) **15** minutes to **7**
   (c) **15** minutes past **7**
   (d) **25** minutes to **8**

3.

Chapter 3 – Time Intervals

Objective

◆ Use a clock to find the time interval in minutes.
◆ Use a clock to find the time interval in hours.
◆ Find the end time when given the start time and the minutes.
◆ Find the end time when given the start time and the hours.
◆ Recognize time intervals that go from a.m. to p.m. or from p.m. to a.m.

Notes

In this chapter your student will learn how to determine the duration of a time interval in either hours or minutes, and how to find when an activity ends given the start time and the amount of time that has passed in minutes or in hours.

At this level the time interval will be minutes or hours, not both. Your student will be able to find the elapsed time or the end time by counting up hours, or counting up 5-minute intervals. The problems will include situations where the time changes from one hour to the next even when the elapsed time is in minutes, such as from 10:40 to 11:10. Your student can count up by 5 minutes, and needs to remember to go from 55 to the hour to 0 minutes and then 5 after the hour. At this level, allow her to use a clock face to help her determine the time interval or the ending time. After a while, she will be able to count up by minutes or hours from the start time without the help of a clock face.

In *Primary Mathematics* 3 students will learn to find the time interval without a clock face, and learn how to find the time interval for hours and minutes.

In some problems, the end time will go from a.m. to p.m. or vice versa compared to the start time.

For the activities in the lessons on the next few pages of this guide, it will be helpful if you have two clocks and can set the start time on both clocks and use the second clock to move the minute hand to the end time so that your student can compare the times. If you have only one clock, put your finger at the start time and keep it there as you move the minute hand to the end time.

Material

◆ Two analog clocks with geared hands
◆ Large demonstration clock, if available

(1) Find the time interval in minutes or hours

Activity

Choose a familiar activity with a time interval of less than an hour and where the start and the end time are in the same hour. Use times where the minutes are multiples of 5. Write down the start time and the end time. Show the start time on the clock. Have your student move the minute hand and count by fives until the end time. Ask him how long the activity lasts.

Do a few other examples.

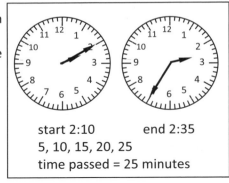

start 2:10   end 2:35
5, 10, 15, 20, 25
time passed = 25 minutes

Discussion

Concept page 80

Task 1, p. 81

1. (a) 25 min
   (b) 15 min
   (c) 25 min
   (d) 30 min

Activity

Show a start time on a clock of 4:00. Write the time. Tell your student that this is the time that some activity starts, e.g. a movie. Ask her to move the minute hand all the way around once and tell you the new time. Tell her that this is the time part way through the movie. Ask her how much time has passed. (60 minutes or 1 hour.) Remind her that 60 minutes is 1 hour. Point out that the minutes are the same on both the start time and the end time, but the hours have changed. Ask her to move the minute hand around one more time and tell you the new time. Write it down. Tell her that this is when the movie ended, and ask her long the movie lasted.

Set the time for 2:30 and tell your student that this is another start time for a different movie. Move the minute hand around once as he counts by fives until he gets to 60. Ask him what time it shows (3:30) and how much time the movie has lasted so far. Point out that again the minutes have not changed between the start and end time, just the hours. Ask him to move the minute hand all the way around two more times and tell him that this is when the movie ended. Write the time and point out that the minute hand is at the same place it was it was for 2:30, and the hour hand has moved forward 3 hours, from between 2 and 3 to between 5 and 6. So this movie lasted 3 hours. The minutes on both the start time and the end time are the same. So all we have to do is look at the hours to find how much time has passed.

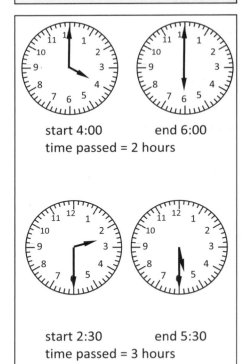

start 4:00   end 6:00
time passed = 2 hours

start 2:30   end 5:30
time passed = 3 hours

Write down a start time of 11:00 and an end time of 3:00. Tell your student that these are the start and end time of some activity, such as a drive between towns. Ask him how long the drive took. He needs to count up from 11 to 3 by switching to 1 after 12. You can show this switch on a clock.

start 11:00   end 3:00

(11) 12, 1, 2, 3

4 hours

Set the start time at 2:45 and write the time. Move the minute hand to 3:10 as your student counts by 5's. Ask your student for the time and write it down. Point out that in the written time the hours have changed, but a whole hour has not passed. This is because the hour changes every time the minutes get to 60. You might want to draw an analogy to counting up from 28 to 32. We have only counted on 4, but the tens have changed.

start 2:45   end 3:10

5, 10, 15, 20, 25

time passed = 25 minutes

Discussion

Tasks 2-4, pp. 82-83

Workbook

Exercise 3, pp. 122-125 (Answers p. 105)

Reinforcement

Give your student practical experience determining elapsed time in minutes or hours throughout the day.

2. 60

3. (a) 25 minutes
   (b) 6 hours

4. 30

(2) Find the end time from the start time and the time interval

Activity

Set the time on a clock for 10:00 and tell your student it is 10:00 a.m. Ask her what time it will be in 4 hours. It will be 2:00 p.m. Remind your student that the times from 12:00 midnight to 12:00 noon are a.m. (part of the night and morning), and the times from 12:00 noon to 12:00 midnight are p.m. (afternoon and evening and part of the night).

Set the time for 11:20 and write the start time as 11:20 p.m. Ask your student to tell you what time it will be 5 hours later. He needs to count up from 11:20, and remember that by passing 12:00 the time goes from p.m. to a.m. In 5 hours, the time will be 4:20 a.m.

Set the time again for 11:50 and write the time as 11:50 a.m. You can tell her that lunch started at this time and ended a half hour later. Ask her to find what time it is a half hour after 11:50 a.m. and to show the time on the clock. It is 12:20 p.m.

Point out that in half an hour the minute hand moves to a point opposite from where it started.

Discussion

Tasks 5-6, p. 83

Workbook

Exercise 4, pp. 126-127 (Answers p. 105)

Reinforcement

Give your student practical experience determining when activities will end throughout the day.

Extra Practice, Unit 11, Exercise 3, pp. 161-162

Test

Tests, Unit 11, Chapter 3, A and B, pp. 115-122

Enrichment

You can show your student that he can use the position of the minute hand after a half of an hour to quickly find where it will be 35 minutes later (half an hour plus one 5-minute mark) or 25 minutes later (half an hour less one 5-minute mark).

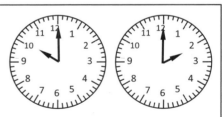

start 10:00 a.m.
11, 12, 1, 2
4 hours later it will be 2:00 p.m.

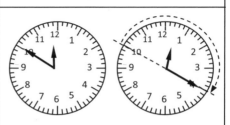

start 11:50 a.m.
1 half hour later it will be 12:20 p.m.

5. 8:35

6. 8:45

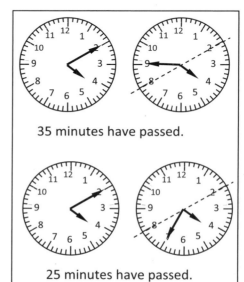

35 minutes have passed.

25 minutes have passed.

Workbook

Exercise 3, pp. 122-125

1. (a) 20
   (b) 15
   (c) 25
   (d) 40

2. (a) 1
   (b) 4
   (c) 3
   (d) 4

3. (a) 15
   (b) 35
   (c) 35
   (d) 50

4. (a) 4:05   5   4:10
   (b) 2:50   10   3:00
   (c) 9:30   40   10:10
   (d) 5:45   7   12:45

Exercise 4, pp. 126-127

1. (a) 4:35 a.m.
   (b) 11:50 p.m.
   (c) 6:15 a.m.
   (d) 6:00 p.m.
   (e) 3:15 a.m.

2. (a) 5:25 p.m.
   (b) 4:15 a.m.
   (c) 1:40 p.m.
   (d) 12:10 a.m.
   (e) 2:55 a.m.

Chapter 4 – Other Units of Time

Objectives

♦ Understand the relationships of various units of time.
♦ Use different units to measure time.
♦ Understand when a specific unit of time should be used to measure the duration of an activity.

Notes

In this chapter your student will learn how to relate other units of time: days, weeks, months, and years. They will learn or review the following relationships:

⇒ 1 day = 24 hours

⇒ 1 week = 7 days

⇒ 1 month = about 4 weeks (28-31 days)

⇒ 1 year = 12 months

⇒ 1 year = 52 weeks

⇒ 1 year = 365 days

Your student will do some simple conversions between units of time using small numbers, such as finding the number of months in 3 years. Although multiplication of a 2-digit number by a 1-digit number has not been covered, your student can easily find the answer with addition: 12 + 12 + 12. Conversions between units of time will be taught more formally in *Primary Mathematics* 3.

Your student will also identify the correct units of time for specific events, such as taking hours rather than weeks to clean the house.

Students learned to use a calendar in Kindergarten. Though the calendar was not specifically taught in *Primary Mathematics* 1, your student should have some idea of its use from daily-life. If not, spend some extra time explaining the function and use of a calendar, and refer to it frequently as the year progresses. If your student does not yet know the days of the week and the months of the year and their sequence, spend some time over the year teaching that, not necessarily just during a math lesson. There are a variety of poems and songs that you can find on the internet to help your student remember the days of the weeks, months of the year, or number of days in each month.

Seconds will be covered in *Primary Mathematics* 3, but you can tell your student that seconds are a shorter unit of measuring time than minutes, and that there are 60 seconds in a minute.

Material

♦ Calendar

(1) Understand other units of time

Activity

Ask your student for some other ways of measuring time. We can use seconds, days, weeks, and so on. As you proceed with a discussion of these units of time write down the relationship between the units.

Ask your student for the number of hours in a day and the number of days in a week. If one day is 24 hours, how many hours are there in 2 days? (24 + 24 = 48 hours).

1 day = 24 hours
1 week = 7 days
1 month = about 4 weeks
1 month = about 30 days
1 year = 12 months
1 year = 52 weeks
1 year = 365 days

Show your student a calendar and explain its organization. One row is a week. Discuss the names of the days of the week and ask your student how many days there are in a week. Then ask your student how many days there are in 2, 3, 4, 5 and 10 weeks.

Use the calendar to show how the days progress from one month to the next; the number of the day starts over at the beginning of the next month and the first day of the next month is the day of the week after the last day of the previous month.

Tell your student there are 12 months in the year and discuss the months of the year, their names, what season they fall in, and particular events or holidays in specific months and then occur again a year later. Just as the hours of the day cycle around to start over at midnight, so do the days of the week cycle around to start over on Sunday, and the months of the year cycle around to start over on January.

Have your student use the calendar to examine the number of days in each month. You may want to tell him that usually February has 28 days, but every 4 years (leap year) it has 29 days. The rest of the months have 30 or 31 one days. Ask your student about how many days are in 2 months (30 + 30 = 60 days). Ask your student if a month is shorter or longer than 4 weeks. It is longer, or the same for February in non-leap years.

Tell your student that there are 52 weeks in a year and 365 days in a year, except for leap year, when there are 366 days.

Discussion

Concept page 84

Tasks 1-2, p. 85

Workbook

Exercise 5, p. 128 (Answers p. 109)

Reinforcement

Get your student to be aware of and keep track of the time it takes for various activities, from hours to years.

Extra Practice, Unit 11, Exercise 4, pp. 163-164

Test

Tests, Unit 11, Chapter 4, A and B, pp. 123-129

7 days in a week
12 months in a year
31 days this month

1. (a) hours
   (b) days/weeks/months
   (c) years
   (d) weeks
   (e) months
   (f) hours
   (g) months/years

2. 2 days
   15 days
   15 weeks
   2 years

Review 11

Review

Review 11, pp. 86-87

Workbook

Review 12, pp. 129-133 (Answers p. 109)

Test

Tests, Units 1-11, Cumulative, A and B, pp. 129-140

Enrichment

One easy way to determine the number of days in a month is to use your fists (but not to knock it into your student's head...). You can teach the following to your student. Hold out your two fists in front of you. There are knuckles and valleys between each pair of knuckles. Assign the months in order to each knuckle and valley from left to right. If a month falls on a knuckle, it has 31 days, if it falls on a valley, it has 30 days, except for February.

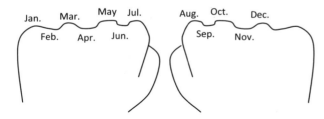

1. (a) 92	(b) 671 (c) 178
2. (a) 371	(b) 65 (c) 425
3. (a) 25	(b) 24 (c) 90
4. (a) 9	(b) 9 (c) 5
5. (a) $9	(b) $1.50 (c) $0.45

6. 40 min

7. 10:10 a.m.

8. (a) minutes
  (b) hours
  (c) hours
  (d) minutes
  (e) hours
  (f) months
  (g) days

9. $\dfrac{1}{7}$

10. $\dfrac{5}{8}$

11. 4

12. (a) $358 - 169 = 189$
    There were **189** men.
  (b) $189 - 169 = 20$
    There were **20** more men.

13. (a) $1.80 + $7.95 = $9.75
    He spent **$9.75**.
  (b) $10 - $9.75 = $0.25
    He got **$0.25** (one quarter) change.

Workbook

Exercise 5, p. 128

1. 12 months — 1 year
   7 days — 1 day
   52 weeks — 1 week
   24 hours — 1 year

2. (a) longer
   (b) longer
   (c) longer
   (d) shorter

3. 12 + 12 + 12 = 36 (or 12 x 3 = 36
   36 months

4. 7 + 7 + 7 + 7 + 7 + 7 = 42
   42 days

Review 12, pp. 129-133

1. (a) 569
   (b) 700
   (c) 444
   (d) 300

2. (a) 40
   (b) 16
   (c) 826
   (d) 704

3.

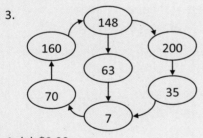

4. (a) $9.00
   (b) $6.40
   (c) $3.35
   (d) $3.10

5. (a) <   (b) >
   (c) =   (d) >

6. (a) 36   (b) 9
   (c) 40   (d) 8

7. (a)   6   9   12 15 18
   (b) 1   2   3   4   5   6

8. (a) hours
   (b) months
   (c) years
   (d) days/months

9. There are 6 pieces, so each piece is $\frac{1}{6}$ of the whole.

10. 24 ÷ 4 = 6
    There were 6 groups of 4.
    Each group of 4 is one out of 6 groups.
    A group of 4 is $\frac{1}{6}$ of the class.

11. 2

12. (a) $6.80 + $5.60 = $12.40
       He spent $**12.40**.
    (b) $12.40 + $15 = $27.40
       He had $**27.40** at first.

13. 1 table: 5 chairs
    9 tables: 5 x 9 = 45 chairs
    He bought **45** chairs.

14. 22 ÷ 4 = 5 with 2 left over.
    Sandra borrowed **2** books.

15. 38 + 298 + 162 = 498
    498 people went jogging.

Unit 12 – Capacity

Chapter 1 – Comparing Capacity

Objectives

♦ Compare capacity.

Vocabulary

♦ Container

Notes

The capacity of a container is how much it will hold. Your student will learn this term in the next chapter.

In Prim*ary Mathematics* 1A students compared the capacity of two or more containers and measured the capacity with non-standard units. In this chapter, students will review various ways of comparing the capacity of two or more containers.

There are several ways to compare capacity.

⇒ Visual inspection of containers. This is not a very accurate method, since the difference in capacity needs to be quite obvious, or the containers need to be similar. A shorter, wider container could have a greater capacity than a taller, narrower container, and your student might assume it has less capacity just because it is shorter.

⇒ Fill up one container, and then pour the water from it into the other. If there is some remaining in the first when the second is full, the first has a greater capacity. If the first can be emptied into the second and not fill it, the second has greater capacity (textbook p. 88).

⇒ Fill up each container, and pour the contents into smaller containers of equal capacity to see how many more small containers one fills than the other (task 1, textbook p. 89).

⇒ Fill up both containers and pour the contents into two larger containers of equal size and shape, or one container and mark the level on each, and compare the levels.

Material

♦ Various containers of different capacity
♦ Two containers of similar capacity, but not equal
♦ Paper cups or cups of the same size and shape

(1) Compare the capacity of containers

Activity

Show your student two of the containers with similar capacity and ask him which holds more water. Use containers where he cannot just look at them and be sure they have the same capacity – one might be taller and thinner than the other. Ask him how he can be sure of his answer. Allow him to demonstrate his ideas.

Put different amounts of water into 3-4 cups of the same shape and size. Ask your student how she knows which one has the largest amount. She can tell by the level of the water.

Discussion

Concept page 88

Tasks 1-2, p. 89

> 1. jug: 5 glasses
>   Bottle: 3 glasses
>
> 2. B holds the most.
>   A holds the least.

These pages show several ways to compare capacity. The answer to task 2 can be deduced from the previous task, as well as by the size of the containers.

Discuss another method for comparing capacity, if your student has not already suggested it. She can fill up the two containers and then pour the water from each into two larger containers of the same size and compare the level of water.

Activity

Use 3-4 containers of different shape. Have your student fill the smallest one completely with water, and then pour that water into another container. Repeat for all the containers, and then fill the first again. Ask him which has the most water. (They all have the same amount.) Ask him whether the level of water is the same. (No.) Discuss why the levels are not the same, even though they have the same amount of water.

Workbook

Exercise 1, pp. 134-138 (Answers p. 116)

Reinforcement

Extra Practice, Unit 12, Exercise 1, pp. 169-170

Test

Tests, Unit 12, Chapter 1, A and B, pp. 141-146

Chapter 2 – Liters

Objectives

- Understand the liter as a unit of measurement.
- Measure the capacity of containers in liters.
- Estimate the capacity of containers in liters.
- Solve word problems involving liters.

Vocabulary

- Liter
- Capacity
- Liquid

Notes

In this chapter the liter is introduced as a standard unit of measurement. Your student should get a feel for how much a liter is and be able to estimate the capacity of containers in liters; that is, whether a container holds less than a liter or more than a liter, and if more, whether a few liters or many liters.

For your information, the SI unit (international standard unit of measurement) for length is a meter, and for mass is a kilogram. Students learned about these units of measurement in *Primary Mathematics* 2A. The SI for volume is a cubic meter; that is, the amount of space equivalent to a cube that is a meter long on each side. Most countries, however, measure volume in liters. One liter is equal to a cubic decimeter, or 1,000 cubic centimeters, and there are 1,000 liters in a cubic meter. A liter is a little more than a quart (1.057 U.S. liquid quarts).

In the textbook the liter is abbreviated with a cursive ℓ. It is usually abbreviated with a capital L. Your student can use a capital L, rather than trying to write the cursive ℓ.

Material

- Liter measuring cup
- Paper cups or cups of the same size and shape
- Various containers

(1) Measure capacity in liters

Activity

Concept page 90

Tell your student that most countries measure water or other liquids (e.g. milk, oil, gasoline) in liters. Show him a 1-liter beaker or measuring cup. If you are using a measuring cup that has quarts on one side, point out which side shows a mark for 1 liter. Show him how to fill the measuring cup or beaker to the line marked as 1 liter.

The textbook says that we write ℓ for liters. Tell your student that **L** is also commonly used to stand for liters.

Do the second activity on this page, or have your student use a cup she commonly drinks from, fill it with water, pour it into the liter measuring cup, and repeat. Count the number of times she pours her cup into the liter container. This can help her get a feel for how much a liter is when she does not have a liter measuring cup to look at.

Give your student various containers and ask him to guess if they contain more or less than one liter, and then find out using the measuring cup.

Discussion

Tasks 1-4, pp. 91-93

> 1. The **jug** holds the most.
>   The **glass** holds the least.
>
> 2. 1 liter
>
> 3. **B** can hold more.
>   It can hold **3 liters** more.
>
> 4. 4

Activity

Task 5, pp. 93-94

Rather than marking a bottle, you can just use the measuring cup. Provide your student with several containers and a chart similar to the one on p. 94. Have your student estimate and then measure how much each container holds.

Workbook

Exercise 2, pp. 139-140 (Answers p. 116)

Reinforcement

Extra Practice, Unit 12, Exercise 2, pp. 171-172

Test

Tests, Unit 12, Chapter 2, A and B, pp. 147-152

Chapter 3 – Gallons, Quarts, Pints and Cups

Objectives

♦ Understand the gallon, quart, pint, and cup as units of measurement.
♦ Measure the capacity of containers in gallons, quarts, pints, and cups.
♦ Estimate the capacity of containers in gallons, quarts, pints, and cups.
♦ Compare gallons, quarts, pints, and cups and do simple conversions.
♦ Compare quart to liter.

Vocabulary

♦ Gallon
♦ Half-gallon
♦ Quart
♦ Pint
♦ Cup

Notes

This chapter introduces gallons, half-gallons, quarts, pints, and cups as units of measurement. They learn the abbreviations for these units of measure and some conversion units (e.g. there are 2 cups in a pint). Converting between these measures will be covered more thoroughly in *Primary Mathematics* 3B.

2 cups = 1 pint
2 pints = 1 quart
2 quarts = 1 half-gallon
2 half-gallons = 1 gallon
4 quarts = 1 gallon

Your student should be able to estimate whether a container holds one or more cups, quarts, or gallons, and which unit would be appropriate for finding the capacity of a container. For example, a bathtub would have a capacity in gallons, but a coffee mug would have a capacity in cups.

Be sure your student understands that he is learning two systems of measurement, one is used in the U.S. and one, the metric system, is used in most of the rest of the world and in science. You may want to review other units of measurement.

Meters, centimeters, kilograms, grams, and liters are part of the metric system.

Inches, feet, yards, pounds, ounces, gallons, quarts, pints, and cups are customarily used in the U.S. and variously called customary units of measurement, standard units of measurement, or Imperial units of measurement.

The metric system is used in the U.S. in science.

Material

♦ Quart measuring cup
♦ Pint or cup measuring cup
♦ Half-gallon container
♦ Gallon container
♦ Various other containers of items that commonly come in pints or quarts, e.g. milk products

(1) Measure capacity in U.S. customary units

Activity

Have your student look at various containers you might have where the capacity is marked in cups, pints, quarts, half-gallons, or gallons. Tell your student the capacities and that these are used to measure the amount of liquids in the U.S.

Have your student look at the quart measuring cup (which is usually also marked in liters on the other side) and point out the cup and pint levels. Help your student fill the containers to the cup or pint or quart levels. Fill the quart measuring cup to the quart level and have your student look at the markings on the other side to see that the level is not quite at the liter mark. A quart is a little less than a liter. If you want, you can have her use the smaller measuring cup and containers such as milk jugs or containers, to find out how many cups are in a pint, pints in a quart, quarts in a half-gallon, and quarts and half-gallons in a gallon.

Discussion

Concept page 95

Tasks 1-6, pp. 96-98

For task 3, you may want to have your student use a measuring cup to find out how many cups are in a pint and in a quart instead.

For task 4, you can just use a pint measuring cup.

Activity

Have your student refer back to p. 95 and help her create a chart similar to the one shown here at the right.

If your student has trouble with this, you can use linking cubes. 1 cube is a cup. Put 2 together to represent a pint. Have him show how many cups are in a quart by putting two pint-sets together. Then put quart-sets together to make a half-gallon, and so on.

Point out to your student that if he remembers that there are 2 cups in a pint, 2 pints in a quart, 2 quarts in a half-gallon, and 2 half-gallons in a gallon, it is easy to figure out the other comparisons by simply multiplying by 2 each time. He has to remember the order gallon → half-gallon → quart → pint → cup.

Workbook

Exercise 3, pp. 141-142 (Answers p. 116)

Reinforcement

Extra Practice, Unit 12, Exercise 3, pp. 173-176

Test

Tests, Unit 12, Chapter 3, A and B, pp. 153-158

---

1. 16 cups

2. The larger one on the left holds more.

3. (b) quart

4. (a) no
   (b) yes

5. (a) gallon
   (b) pint, cup
   (c) cup, pint
   (d) quart

---

1 pint	↔	2 cups
1 quart	↔	2 pints
1 quart	↔	4 cups
1 half-gallon	↔	2 quarts
1 half-gallon	↔	4 pints
1 half-gallon	↔	8 cups
1 gallon	↔	2 half-gallons
1 gallon	↔	4 quarts
1 gallon	↔	8 pints
1 gallon	↔	16 cups

Workbook

Exercise 1, pp. 134-138

1. (a) bottle   (b) pail

2. (a) fish tank   (b) watering can

3. (a) glass   (b) bowl

4. (a) mug   (b) bowl

5. **A** holds more water than **B**.

6. **A** holds less water than **B**.

7. B

8. X

9. **B** holds the most.
   A holds the least.

10. X holds **2** more than Y.
    X holds **2** less than Z.

Exercise 2, pp. 139-140

1. (a) 4
   (b) 4

2. (c) 3
   (b) 3

3. 600 ℓ − 458 ℓ = 142 ℓ
   Station B sold **142 liters** more.

4. 1 bottle: 4 ℓ
   5 bottles: 4 x 5 = 20 ℓ
   He bought **20 liters**.

Exercise 3, pp. 141-142

1. (a) 7
   (b) 7

2. (a) 6
   (b) 6

3. 250 gal − 105 gal = 145 gal
   145 gal more are needed.

4. 16 qt are being grouped by 8.
   16 qt ÷ 8 = 2 qt
   There were **2 qt** of orange juice in each jug.

Review 12

Review

Review 12, pp. 99-100

For problem 16, if your student has not learned the multiplication facts for 4 x 12, see if she can come up with a way of solving this problem. She could convert the amount each child drank into cups and then add to find the total number of cups.

Workbook

Review 13, pp. 143-147 (Answers p. 118)

Problem 5 involves ordering fractions other than unit fractions, but students use a drawing to compare the fractions, and should be able to do it.

Test

Tests, Units 1-12, Cumulative, A and B, pp. 159-170

Enrichment

Ask your student the following:

⇒ How many cups are in 8 pints?   (16 cups)

⇒ How many cups are in 5 quarts?   (20 cups)

⇒ How many quarts are in 6 gallons?   (24 quarts)

⇒ How many quarts are in 14 pints?   (7 quarts)

⇒ How many gallons are in 12 quarts?   (3 gallons)

⇒ How many pints are in half a quart?   (1 pint)

⇒ How many cups are in half a quart?   (2 cups)

⇒ How many pints are in 10 cups?   (5 pints)

1. (a) 825   (b) 600   (c) 800

2. (a) 301   (b) 146   (c) 199

3. (a) 32   (b) 30   (c) 80

4. (a) 6   (b) 10   (c) 9

5. (a) $6.30 (b) $6.01 (c) $1.85

6. $5.80 + $2.75 = $8.55
   The book costs **$8.55**.

7. 11:20 a.m.

8. 20 ℓ – 4 ℓ = 16 ℓ
   He used **16 liters**.

9. 30 ℓ – 12 ℓ = 18 ℓ
   18 liters more are needed.

10. 52 ℓ + 38 ℓ = 90 ℓ
    He sold **90 liters** in 2 weeks.

11. 1 day: 5 ℓ
    7 days: 5 ℓ x 7 = 35 ℓ
    He uses **35 liters** in a week.

12. (a) **A** holds more.
    (b) 12 ℓ – 8 ℓ = 4 ℓ
        It holds **4 liters** more.

13. 14 ÷ 3 = 4 with 1 left over.
    5 bottles are needed.

14. (a) 203 – 128 = 75
        There were **75** children.
    (b) 128 – 75 = 53
        There were **53** more adults than children.

15. 1 gallon = 4 qt
    She drank 3 out of 4 quarts.
    She drank $\frac{3}{4}$ of the milk.

16. (a) 12 quarts
    (b) 1 qt = 4 cups
        12 qt = 4 c x 12 = 48 c
        Or 24 c + 16 c + 8 c = 48 c
        They drank **48 c**.
    (c) Group quarts by 4. 12 ÷ 4 = 3
        They drank **3 gallons**.

Workbook

Review 13, pp. 143-147

1.
8	10	12	14	16	18	20
12	15	18	21	24	27	30
16	20	24	28	32	36	40
20	25	30	35	40	45	50
40	50	60	70	80	90	100

2. (a) ℓ
   (b) min
   (c) m
   (d) g
   (e) kg
   (f) h
   (g) cm
   (h) ℓ

3.
18	40	36	90
9	8	8	10

4. (a) five dollars and ninety cents
   (b) nine dollars and fifty cents
   (c) five dollars and nine cents
   (d) nine dollars and five cents

5. Sections shaded may vary.

$$\frac{2}{3} \ , \ \frac{3}{4} \ , \ \frac{5}{6}$$

6. $\frac{1}{3}$

7. 2 hours

8. 20 to 5

9. (a) a.m.
   (b) p.m.
   (c) p.m.
   (d) a.m.

10. 10 months

11. money in quarters = $2
    money in dimes = $1.20
    He has **$3.20**.

12. $10 − $3.95 = $6.05
    She received **$6.05** in change.

13. 5 tickets: $35
    1 ticket: $35 ÷ 5 = $7
    1 ticket cost **$7**.

14. 210 − 145 = 65
    He read **65** pages the second day.

15. 1 week: $4
    6 weeks: $4 x 6 = $24
    He can save **$24** in 6 weeks.

16. $245 + $65 = $310
    His sister saved **$310**.

Unit 13 – Tables and Graphs

Chapter 1 – Picture Graphs

Objectives

- Review tally charts, tables, and picture graphs.
- Create tables.
- Create picture graphs with a scale.
- Interpret picture graphs with a scale.
- Answer questions using information from a picture graph.

Vocabulary

- Tally chart
- Table
- Picture graph

Notes

A graph is a pictorial representation of data. It offers a visual display of relationships, making them more noticeable than if the data were presented simply in numerical form. There are a variety of types of graphs, including picture graphs, bar graphs, and line graphs.

Tallies are an easy way to keep track of things as you are counting them; they are not a particularly helpful way of presenting data. Once the data have been collected, they can be organized into a table that makes it easy to find and compare specific pieces of data.

Picture graphs are an eye-catching way of presenting the data in a way that makes it easy to compare data. A single glance at a bar graph can tell you how quantities compare.

A bar graph uses the length of solid bars to represent numbers and compare data.

Students learned how to make simple tally charts and to create, read, and interpret simple picture graphs and bar graphs in *Primary Mathematics* 1B, where each picture, symbol, or square stood for 1 item. In this chapter they will review tally charts and picture graphs, and then will learn to create and interpret picture graphs where the symbol represents more than one item. At this level, the scale will be 2, 3, 4, 5, or 10 so that students can use the multiplication and division facts for those numbers to interpret the graphs.

Material

- Appendix pp. a32-35
- Counters
- Multilink cubes

(1) Understand scaled picture graphs

Discussion

Concept pages 101-103

> Have your student look at p. 101. Remind him that one way to count items that are mixed up is to make a tally chart. To make a tally chart, we make a mark for each object, and then a mark through 4 of them for the fifth object. Appendix page a32 has a picture similar to the one shown in the text. Get your student to tally the fruit. Tell him that when he makes a tally chart, he does not have to do only one kind of fruit at a time, but could start at one side of the picture, check off a fruit with a little mark to show it is counted, make a tally mark in the correct place on the chart, check off another fruit and make a tally mark and so on. Then have him count his tally marks and write the actual number of each kind of fruit in the third column.

4 types
12 mangoes
6 pears
8 apples
4 oranges
30 total

(a) 2 fruits
(b) 12 fruits
(c) 2
(d) 4
(e) mango
(f) orange

After he has finished, tell him that in this case it is probably easy enough to simply count each kind of fruit one by one rather than making a tally chart. But sometimes we might want to collect information where we can't do that, such as how many hybrid cars pass by versus trucks. Then, it is easier to make a mark for each vehicle than try to keep track of them by counting.

Tell your student that once we have tallied and written how much of each kind we have, we can draw a *picture graph* so that it is easier to compare how many we have of each kind. Have your student look at p. 102. Each circle stands for one fruit.

Ask your student what we could do if we had a lot more fruit. If we had 100 mangoes instead of 12 we could not fit 100 round circles on a page. Ask for suggestions on how we could show a picture graph on a single page if we had more fruit.

Have your student look at p. 103. Here, triangles stand for the fruit, but each triangle now stands for two fruits. If each symbol stands for more than one fruit, we don't need as many pictures but we can still easily see which type we have the most of and which we have the least of.

Task 1, p. 104

Workbook

Exercise 1, pp. 148-151 (Answers p. 122)

1. (a) 8
   (b) 24
   (c) Carlos
   (d) 12
   (e) 16

Reinforcement

Use appendix p. a 33. Give your student four colors of counters or four categories of other objects (e.g. buttons with different shapes). The amount of each category should be a multiple of 5, and no more than 50. Have your student count the number in each category, decide what one circle on the graph should stand for, label the graph with the categories, and color the appropriate number of circles. Have your student then use the graph to answer questions similar to those in the text, such as, which color there is the most of, how many more there are of one than another, and so on.

(2) Interpret picture graphs

Activity

Give your student 6 multilink cubes. Ask her, if 1 cube stands for 3, how many 6 cubes stand for. Then give her 1 cube. Tell her it stands for 4, and ask her how many cubes she would use to stand for 24.

Now give your student 5 cubes. Tell him they stand for 25. Ask him what 1 cube stands for. Then have him show you how many would be used to stand for 40. Have him line them up and ask him how many more are represented by the row of 8 than by the row of 5. He can find the answer by subtracting 25 from 40, or by subtracting 5 from 8, and then multiplying by 5. If the difference in number of cubes is 3, then the difference in amount they represent is 3 x 5 = 15.

Discussion

Tasks 2-3, pp. 105-106

> Point out the table on p. 105. Tell your student that a table is another way of showing the information. As she goes through the tasks, ask her whether it was easier to find the answer from the table or from the graph. Graphs are easier for visually comparing data, but tables have the exact values.
>
> 2(d): You can also ask how many stars would now be in the zoo column.
>
> 3(c): Students have not yet multiplied 25 by 2 but do know that multiplying is the same as doubling. You can ask your student what he would get if each sold for $1. He would get $25. If each sold for $2 instead, he would get double that, or $25 + $25 = $50.
>
> 3(d): Since he sold 10 swordtails, this is just $30 ÷ 10 = $3.

2. Each star stands for 3 children (a) 5 (b) zoo (c) 9 (d) 2 3. Guppy: 25 Goldfish: 25 Tiger barb: 5 Swordtail 10 (a) goldfish (b) 35 (c) 50 (d) 3

Workbook

Exercise 2. pp. 152-155 (Answers p. 122)

Reinforcement

Have your student complete the graph on appendix p. a34 using the information given on that page.

Extra Practice, Unit 13, Exercise 1, pp. 181-184

Test

Tests, Unit 13, Chapter 1, A and B, pp. 171-174

Enrichment

Have your student do the problems on appendix p. 35. The last 4 are more challenging than the first 5. The answers are given at the right.

1. 16
2. 8
3. 5
4. 4
5. 50
6. 9
7. 8
8. 8
9. 6

Workbook

Exercise 1, pp. 148-151

1. (a) 3
   (b) 6
   (c) 2
   (d) red
   (e) yellow

2. (a) 8
   (b) 5
   (c) 5
   (d) 3
   (e) Cameron

3. (a) 10
   (b) 14
   (c) 4
   (d) 6
   (e) 6

4.

		●	
	●	●	
	●	●	
	●	●	●
●	●	●	●
●	●	●	●
Divya	Sally	Weilin	Rosni

Exercise 2, pp. 152-155

1. (a) 10
   (b) 30
   (c) 12

2. (a)

   (b) △ △ △ △ △ △ △ △ △ △

3.

David's savings	
March	$9
April	$3
May	$12
June	$6

   (a) $9
   (b) $30
   (c) $12
   (d) $3

4. (a) 50
   (b) Matthew
   (c) Annie
   (d) 10
   (e) 30
   (f) 70

5. (a) no
   (b) yes
   (c) no
   (d) yes
   (d) no

Chapter 2 – Bar Graphs

Objectives

- Interpret bar graphs.
- Answer questions using information from a bar graph.
- Create simple bar graphs.
- Find the range and mode of data.

Vocabulary

- Bar graph
- Scale

Notes

In *Primary Mathematics* 1B students learned to interpret simple bar graphs where one square or rectangle on the graph represented one item. The bar graphs were essentially picture graphs but with adjacent squares or rectangles instead of pictures.

In this chapter your student will learn to interpret bar graphs with scales along one side of the graph. The bars can be vertical or horizontal. The scale is the subdivision of each axis (horizontal and vertical) side. The scale may be numerical or categorical. In bar graphs, the scale is numerical on one side and categorical on the other.

You may wish to have your student collect information and create graphs to present data. Generally, at this level, students are provided with some task that dictates a fairly simple graph, such as the number of each color candy in a bag of candy, or recording the weather, or finding out which flavor of something or color people like best, with limited choices. At this level, students do not really have the mathematical background to deal with actual data, which is usually messier than data collected under controlled or contrived situations. Graphs in math textbooks are going to be much simpler than graphs from data from the "real world." If you are going to have your student collect and present data, try to come up with something he is actually interested in, perhaps through a different subject such as science, and give him assistance in creating the axes (sides) of the bar graphs. Students will learn to use graph paper and choose suitable scales to create bar graphs for data in *Primary Mathematics* 3. The focus at this level is on reading and interpreting the graphs and understanding the scale, not creating bar graphs, which will be easier for students when they are older.

Bar graphs and picture graphs represent the same information or data. They are different only in the manner in which the data are presented. Picture graphs are more eye-catching. Bar graphs with scales are more useful when representing information with larger numbers.

Students will be finding the *range* (difference between lowest to highest value) and *mode* (the value that appears most often in a set of data). They do not have to learn these terms at this level. They will learn the terms median and mode in *Primary Mathematics* 4 and 5, and the term mean in *Primary Mathematics* 5, and examine what each of these tells them about the data at that time.

Material

- Appendix p. a36

(1) Interpret bar graphs

Discussion

Concept pages 107-108

Tell your student that the bar graph represents the same data as the picture graph. Instead of saying how many each square stands for, a **scale** is used. Point out the scale on the left, and guide your student in determining what each unlabeled mark stands for. The interval between each mark is 1, but only every other mark is labeled. Note that the height of the green bar goes to half way between the labeled marks. Then have your student answer the questions. Note that the **he** in (b) is the answer to (a) and the **he** in (d) is the answer to (c).

(a) Tyrone
(b) 9
(c) Sam
(d) 2
(e) 7
(f) Matthew: 4
Pablo: 6
(g) 21
(h) 2

Task 1, p. 109

Ask your student to determine what each interval between marks represents and the value of the unlabeled marks on the vertical scale. Point out that when the numbers get larger, there would not be enough room for the graph on a page if each mark was for every 1 more. Extra marks are added (one for every 5 rather than for every 10) so that it is easier to read the values for the bars, for example the first bar. For questions (e) and (h), discuss why the table or the graph might be used. Both could be used to answer the same question, such as finding the highest score using the graph and then finding the value using the table.

1. (a) 80
(b) Math
Social Studies
(c) 5
(d) Science
(e) Answers will vary.
(f) 90
(g) 75
(h) Answers will vary.

Task 2, p. 110

Point out that bar graphs, like picture graphs, can be drawn with the bars going sideways rather than up and down.

Workbook

Exercise 3, pp. 156-158 (Answers p. 126)

2. (a) $30
(b) $30
(c) July
(d) June
(e) June
(f) $30 + $60 + $15 + $30 = $135
She saved **$135.**
(g) May and August
(h) $30 + $30 = $60
She saved **$60** in May and
August.

(2) Interpret tables and bar graphs

Discussion

Task 3, p. 111

After your student has answered the questions, ask him if any of the them might have been answered more easily with a graph. Have him create a graph for the data in the table using appendix p. a36.

Task 4, p. 112

You can either have your student create a graph for this data using appendix p. a36 before answering the questions, or have him create the graph after answering them, and then going back over the questions to see if a bar graph makes a difference. Many of these questions are more easily answered using a bar graph and there will be an obvious benefit to using a bar graph in situations such as this involving frequency data.

Task 5, p. 113

Discuss the scale for this graph. Ask your student for the interval between each mark. (10.) Point out that graphs are usually drawn on graph paper, such as the one on appendix p. a37. When they are not, the dotted line from the top of the bar to the scale make it easier to find the value for each bar.

3. (a) 60
   (b) 75
   (c) $75 - 50 = 25$
   (d) Baljit
   (e) David
   (f) Maria

4. (a) 1
   (b) 2
   (c) 5
   (d) 2
   (e) 20
   (f) 10
   (g) 25
   (h) 15

5. (a) 320
   (b) $440 - 250 = 190$
      190 more people visited on Friday than on Thursday.
   (c) Wednesday
   (d) Thursday
   (e) Tuesday ($160 + 160 = 320$)
   (f) $320 - 200 = 120$
      There were **120** children.

Workbook

Exercise 4, pp. 159-161 (Answers p. 126)

Reinforcement

Extra Practice, Unit 13, Exercise 2, pp. 185-186

Test

Tests, Unit 13, Chapter 2, A and B, pp. 175-178

Workbook

Exercise 3, pp. 156-158

1. (a) 10
   (b) 6:45 a.m.
   (c) 4
   (d) 6:30 a.m.
   (e) 18

2.

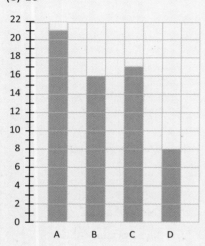

   (a) 4
   (b) 2
   (c) 62

3. (a) 500
   (b) 150
   (c) 650
   (d) 400

Exercise 4, pp. 159-161

1. (a) 12
   (b) Mary
       18
   (c) Wendy
       9
   (d) Wendy

2. (a) 45
   (b) 10
   (c) Ryan
   (e) 60 + 70 + 55 = **185**

3. (a) 180
   (b) 420 − 300 = 120
   (c) Samy
   (d) 100
   (e) 420
   (f) 420 − 100 = **320**

Review 13

Review

Review 13, pp. 114-115

In the first printing of the textbook, 15(b) is a 2-step problem. Your student may need help determining the first step. (Total number of cars and vans.)

15(c) uses the term difference which has not been formally taught, but could be interpreted from the lengths of the bars. You can tell your student to find which type has the greatest number and how much, which has the least type and how much, and how many more cars there are than buses.

Workbook

Review 14, pp. 162-167

5(a): This problem involves comparing fractions that are not unit fractions, but a diagram is supplied so students can concretely compare the fractions.

9(c): This is a 2-step problem.

15(b): If your student does not get $\frac{1}{3}$ as the answer, have her act out the problem with counters.

Test

Tests, Units 1-13, Cumulative, A and B, pp. 179-186

1. (a) 204   (b) 1000   (c) 510

2. (a) 109   (b) 208   (c) 48

3. (a) 18   (b) 12   (c) 40

4. (a) 4   (b) 4   (c) 7

5. How many 4's in 28?
   $28 \div 4 = 7$
   He got **7** pieces.

6. $250 ℓ - 185 ℓ = 65 ℓ$
   65 ℓ more are needed.

7. $30 - 14 = 16$
   He has **16** mangoes left.

8. 5:20 p.m.

9. $\$9 - \$3.80 = \$5.20$
   The book cost **\$5.20**.

10. (a) 1 day: 1 orange
        5 days: $2 \times 5 = 10$ oranges
        He eats **10** oranges in 5 days.
    (b) 7 days: $2 \times 7 = 14$ oranges
        He eats **14** oranges in a week.

11. (a) $\frac{1}{4}$

    (b) $\frac{1}{2}$

    (c) 45

12. 14 days

13. 24 months

14. (a) less than 6 weeks
    (b) more than 20 days

15. (a) **15** more vans than buses
    (b) $45 + 30 = 75$; $90 - 75 = 15$
        15 spaces were not occupied.
    (c) $45 - 15 = 30$
        There are **30** more cars than buses.

Workbook

Review 14, pp. 162-167

1. (a) 408   (b) 250

2. (a) 950
   (b) 728
   (c) 972
   (d) 620
   (e) 590

3. (a) 59   (b) 42
   (c) 37   (d) 76

4. (a) 150   (b) 300
   (c) 500   (d) 841
   (e) 396   (f) 549
   (g) 625   (h) 73

5. (a) =   (b) <

6. (a) 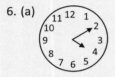   (b)

7. (a) 10   (40 ÷ 4)
   (b) 30   (5 x 6)

8. (a) 120 − 70 = **50**
   (b) 70 + 80 + 30 + 120 = **300**

9. (a) 120 + 85 = 205
       There are **205** girls.
   (b) 120 + 205 = 325
       There are **325** children.
   (c) Number of children who wear glasses:
       19 + 16 = 35
       Number of children who do not wear glasses:
       325 − 35 = 290
       Or: 325 − 19 − 16 = 290
       290 children do not wear glasses.

10. 89 + 91 + 90 = 270
    His total score is **270**.

11. Group muffins by 4.
    32 ÷ 4 = 8
    She had **8** boxes.

12. 1 piece: 5 m
    9 pieces: 5 m x 9 = 45 m
    She bought **45 m** of rope.

13. $8.50 − $6.80 = $1.70
    She needs **$1.70** more.

14. 1 week: 6 qt
    10 weeks: 6 qt x 10 = 60 qt
    They drink **60 qt** in 10 weeks.

15. (a) 24 ÷ 3 = 8
        Each boy gets **8** cards.
    (b) Since there are 3 equal groups, and each boy
        has 1 out of 3 groups, each boy has $\frac{1}{3}$ of the
        cards.

        8 out of 24, or $\frac{8}{24}$, is also a correct answer.

Unit 14 – Geometry

Chapter 1 – Flat and Curved Faces

Objectives

- Identify cubes, prisms, rectangular prisms, pyramids, cylinders, cones, and spheres.
- Classify and sort common solids by faces, and by number of edges and vertices.
- Identify common 2-dimensional shapes on 3-dimensional solids.

Vocabulary

- Flat surface (face)
- Curved surface (face)
- Edge
- Vertex
- Cube
- Prism
- Rectangular Prism
- Pyramid
- Cylinder
- Cone
- Sphere
- Base
- Apex

Notes

In *Primary Mathematics* 1A, students learned to identify the four basic shapes (squares, rectangles, circles, and triangles) count their corners and sides, group them by various attributes, and combine these shapes into new shapes. They also learned to identify these common shapes on the surface of solids, identify flat and curved surfaces on solids, and group solids by various attributes, such as whether they could be stacked, rolled, or slid.

In this chapter your student will review flat and curved surfaces, learn the names of common solids, and learn to distinguish and classify solids by the shape of the surfaces and the number of edges and curved edges and vertices.

Your student will need to be able to identify solids in 2-dimensional pictures.

For most students classifying solids by their shape is an easy topic, but memorizing some of the names is not. This chapter may take only one lesson instead of two, but spend time reviewing the names of the solids as you continue in the textbook and identifying solids in the environment.

There are some potential problems with the definition of the terms *face* and *edge.* These terms are often not well defined in elementary math texts because the definition changes with use; that is, there are common use definitions, and, depending on the branch of mathematics, more formal definitions. In the formal study of geometry, a *polygon* is a closed plane figure with straight sides, also called *edges*. Squares, rectangles, octagons, etc. are polygons. A circle is not. A *polyhedron* is a three-dimensional solid which consists of a collection of polygons, usually joined at their edges. Prisms and pyramids are polyhedrons, but a cylinder is not. For a polyhedron, a *face* is defined as one of the polygon surfaces. If the term *face* is restricted to this most restricted use, then a face is limited to a polygon and so is flat with no curved sides, and the term *edge* is limited to a straight line segment.

In this textbook, the terms *face* and *edge* are used in the broadest sense. A face is analogous to a surface, and the term is used with both curved and flat surfaces. An *edge* is the intersection of two faces, and includes both straight and curved edges. A cone therefore has 2 faces and one edge. A sphere has one face (surface) and 0 edges.

Later, students will be given a slightly more restricted definition of a *face*. In *Primary Mathematics* 3 and 4, students will be told that a *face* is a flat surface and an *edge* is formed where two faces meet, but the term *face* will not be restricted to a polygon, and so an *edge* can be curved. They will learn, though, that the faces of prisms and pyramids are polygons.

In order to be more consistent between this level and later levels of *Primary Mathematics*, always use *curved* or *flat* before the term *face*. You can substitute the term *surface* for *face* to apply to both types of faces.

A *vertex* is where three or more edges meet. The term can also be used for the *apex* of a cone.

When students formally study geometry, they should be old enough to understand that certain definitions apply at certain times, and understand that when discussing the face of a polyhedron it can only be flat and have straight edges. Even the word line will be given a restricted definition when used for geometrical proofs and be distinguished from a line segment. At the second grade level, we can use a broader definition of terms closer to everyday usage. So an edge can be curved when the term is applied to geometric figures in general, rather than just to polyhedrons.

The point of this chapter is to help the student recognize common solid shapes by referring to their surfaces, edges, and corners, not to provide formal definitions used for geometrical proofs based on straight lines and angles.

Material

♦ Models of cubes, prisms, rectangular prisms, pyramids, cylinders, cones, and spheres
♦ Objects shaped like cubes, prisms, rectangular prisms, pyramids, cylinders, cones, and spheres
♦ Opaque bag

(1) Identify flat and curved surfaces

Activity

Show your student various models of the four solid shapes: cylinder, triangular prism, rectangular prism, and cone. Discuss the number of curved and flat surfaces and shapes of the surfaces. Use the term *surface*, rather than *face*. Show her a cylinder and ask her if she can identify what shape the curved surface would have if it were flattened out. You can fold a piece of paper around it to show that the surface is a rolled up rectangle. The shapes will be named in the next lesson, but you can use the names informally now. You might want to use the term "box" for the rectangular prism for now.

If your student is not already aware of it, tell her that a square is a special kind of rectangle where all the sides are equal.

If you have available an assortment of shapes, have your student group them by similarity to the models. Or, you can give your student one model shape at a time and have him find items around the house that resemble it with regard to general shape. You can have your student identify curved and flat surfaces on other solids, even those that don't fit nicely into the category of one of the common solids.

Discussion

Concept pages 116-117

Tell your student that the term "face" by itself will only be used for flat surfaces. So we will look at curved faces or flat faces, which are both surfaces.

Task 1, p. 118

You may want to use the word *surface* instead of *face* to help prevent potential confusion later. Have your student relate the pictures with actual solids. She can use the solids (cylinder, prism, rectangular prism, and cone) to answer the questions.

1. (a)	A
(b)	C
(c)	D
(d)	B

Workbook

Exercise 1, pp. 168-171 (Answers p. 134)

Replace the words *face* and *faces* with *surface* and *surfaces*.

(2) Identify common solids

Discussion

Tasks 2-9, pp. 119-121

Use actual solids as you go through tasks 2-8. Show your student the solid, name it, have him identify the shape of the flat faces and count the curved and flat faces, curved and flat edges, and vertices of each. Tell him that an edge is where two faces meet, and a vertex on most shapes is where three or more faces come together (the apex of a cone is also called a vertex). In discussing the shapes, you can discuss classification of prisms, but it is not necessary at this level to master the categories, i.e. that

2. 8 vertices, 6 flat faces, 12 edges
3. Yes
4. The faces are not all the same.
5. A pyramid has only one base.
6. 3 surfaces, (2 flat faces, 1 curved face), 0 vertices
7. It has a vertex.
8. 1 surface (1 curved face) face, 0 edges
9. (a) A, B (b) C, D, F, G, H (c) E (d) D

prisms can be sorted according to the shape of their base, that one kind of prism is a rectangular prism, and that one kind of rectangular prism is a cube. Things you can discuss about each shape are as follows:

⇒ Cube: 6 flat surfaces, 12 edges, 8 vertices.

All the faces are squares.

⇒ Prism: The one shown in task 3 has 5 flat surfaces, 9 edges, and 6 vertices.

A prism has two ends opposite each other which are flat faces with straight edges. We will call them bases. The bases are exactly the same shape and size - if you took away the middle they would fit on top of each other. With the "middle" the opposite sides do not touch, that is, the bases do not have an edge in common. In the prism in the book, the two bases are triangles. You may want to tell your student that this is a *triangular prism*. The bases of a prism can be any flat shape with straight sides; there could be an octagonal prism, for example. The rest of the faces "around the middle" always have 4 sides. A prism where the bases are both squares and the faces around the middle are also squares is a cube.

⇒ Rectangular prism: 6 flat surfaces, 12 edges, 8 vertices.

This is a prism with rectangles for bases. The bases could be squares (which are also rectangles). So a rectangular prism is a special kind of prism. In the case of a rectangular prism, any side could be taken as the base, since for every side there is an opposite side that is exactly the same, but usually the smallest sides are thought of as the bases. A cube is a special kind of rectangular prism where all the faces are squares.

⇒ Pyramid: The one shown on the left has 5 flat surfaces, 8 edges, 5 vertices. The one shown on the right has 4 flat surfaces, 6 edges, 4 vertices.

A pyramid has one side, the base, which can be any shape with straight edges. All the other sides are triangles that meet at a point, or apex, which is opposite the base. The apex is also called a vertex. If you want, you can tell your student that a *rectangular pyramid* has a rectangle or square for a base, and a *triangular pyramid* has a triangle for a base. Both kinds are shown here. But there could also be, for example, octagonal pyramids. The number of faces and edges is determined by the shape of the base.

⇒ Cylinder: 2 flat surfaces, 1 curved surface, 0 straight edges, 2 curved edges, 0 vertices.

A cylinder has circles at the opposite ends. You may want to wrap a piece of paper around a cylinder to show that the curved face is a rectangle if it could be flattened out. Ask your student whether a cylinder is more like a prism or a pyramid. Since it has two bases, it is more like a prism.

⇒ Cone: 1 flat surface, 1 curved surface, 0 flat edges, 1 curved edge, 1 vertex.

A cone is different from a cylinder in that it has one base and one vertex, rather than 2 faces. The base of a cone is a circle. You can ask your student whether a cone is more like a prism or a pyramid.

⇒ Sphere: 0 flat surfaces, 1 curved surface, 0 edges, 0 vertices.

Use examples to distinguish between a sphere and a more ellipsoid shape, such as a football. If your student has good spatial reasoning, you can ask what shape would be made if a sphere were cut straight through (a circle). A major difference between a sphere and all the other shapes is that there is no edge, either curved or straight.

Activity

Put some models of the solids in an opaque bag and have your student identify the shapes by feel.

Workbook

Exercise 2, pp. 172-173 (Answers p. 134)

Reinforcement

Have your student look for representatives of these shapes in the environment. You may want to ask what a shape may have come from. For example, a banister could have been shaped from a cylinder. A cabinet door was probably shaped from a rectangular prism.

Extra Practice, Unit 14, Exercise 1, pp. 191-192

Test

Tests, Unit 14, Chapter 1, A and B, pp. 187-191

Note: Test A, for 2(b) on p. 188: Insert the word "flat" before faces, if it is not there.
A _____ has no edges and no *flat* faces.

Workbook

Exercise 1, pp. 168-171

1. ball   →   orange
   sandwich   →   tent
   drum   →   can
   block   →   present
   pencil case →   cracker box

2. (a) the box
   (b) the calendar

3. (a) rectangle   (b) circle
   (c) square   (d) triangle
   (e) rectangle   (f) rectangle

4. (a)

Solid	Number of flat surfaces	Number of curved surfaces
A	1	1
B	5	0
C	6	0
D	2	1
E	6	0

   (b) 2
   (c) 2
   (d) 2

Exercise 2, pp. 172-173

1. cube   →   tissue box
   pyramid   →   picture of pyramid
   prism   →   box
   cylinder →   picture of cylinder
   cone   →   holiday hat
   sphere   →   globe

2.

Solid figure	Number of surfaces	Number of edges	Number of vertices
Rectangular prism	6	12	8
pyramid	4	6	4
prism	5	9	6
cylinder	3	2	0
sphere	1	0	0

Chapter 2 – Making Shapes

Objectives

♦ Combine common shapes to make new shapes.
♦ Identify half circles and quarter circles.
♦ Divide a shape into common shapes
♦ Make shapes with straight and curved lines.
♦ Identify and continue a pattern according to 2 attributes (color, shape, size, or orientation).

Vocabulary

♦ Half circle
♦ Quarter circle

Notes

In *Primary Mathematics* 1, students combined basic shapes to form compound shapes. This is extended here to more complex shapes, and to include circles and half-circles. Students will also decompose compound shapes into basic shapes.

In *Primary Mathematics* 1, students learned to identify and continue geometric patterns based on one or two attributes — size, shape, or color. In this chapter, the patterns will also include orientation as an attribute.

The lessons in this chapter are short and not likely to be difficult. You can extend the lessons with activities involving tangrams or pattern blocks. There are a number of sites on the internet or books where you can print out puzzles where students form larger shapes from the pattern blocks or tangrams. There are also sites on the internet that allow students to manipulate shapes online. Spatial reasoning is a useful skill for geometry.

Material

♦ Appendix pp. a38-43
♦ Tangrams, pattern blocks (optional)

(1) Compose and decompose compound shapes

Discussion

Concept page 122

Tasks 1-5, pp. 123-126

The shapes to be cut out and covered are reproduced on the appendix pages, slightly larger, which you can copy and cut out, rather than tracing the figures in the textbook.

Workbook

Exercises 3-5, pp. 174-180 (Answers p. 138)

The appendix has the same shapes in the same size as in Exercise 3, which you can copy and cut out to cover the shapes in the textbook.

Reinforcement

Copy the tangrams in the appendix and cut them out. Mix them up and let your student see if he can put them back together to form a square again.

Have your student form other shapes using pattern blocks or tangrams, if you have some.

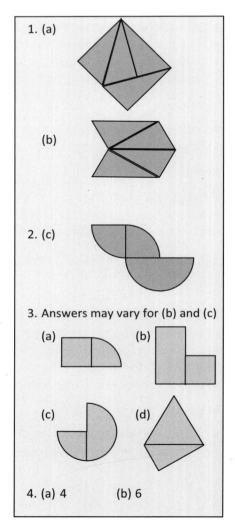

1. (a)

(b)

2. (c)

3. Answers may vary for (b) and (c)
(a)   (b)

(c)   (d)

4. (a) 4   (b) 6

(2) Recognize and extend patterns

Discussion

Tasks 6-7, pp. 127-128

Help your student determine what attribute is changing in each of these. It can be shape, size, color, orientation, or any two of these. She may find it easier to determine the pattern by saying it aloud, and if two attributes are changing to start by looking at each attribute separately. For example, in 6(a) only shape is changing, and the pattern is "circle, triangle, circle, square, circle, triangle, …" so the next shape is a circle. In 7(c), both size and orientation changes, so she could say, "big, little, big, little…" to know that the next shape is big, and then look at the big shape and say, "top, bottom, top…" for the position of the green triangle to determine that the next shape has a green triangle on the bottom.

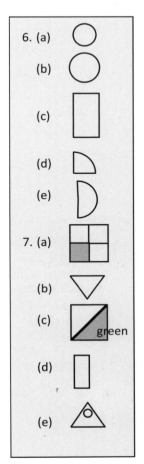

6. (a)
(b)
(c)
(d)
(e)

7. (a)
(b)
(c) green
(d)
(e)

6(a): Shape: circle, triangle, circle, square
6(b): Size: big, small, small
     Shape: circle, circle, square
6(c): Size: large, small
     Shape: triangle, triangle, rectangle, rectangle
6(d): Orientation: upper right, upper right, upper left
6(e): Shape: half-circle, triangle
     Orientation of half-circle: left, right
7(a): Orientation: upper right, lower right, lower left, upper left
7(b): Orientation: up, right, down, left
7(c): Size: big, small (each with different diagonal). Next is big.
     Color of each half. For big square: top green, bottom green
7(d): Shape: rectangle, triangle, triangle
     Orientation: for rectangle: up and down, sideways.
     Orientation for triangle, down, up.
7(e): color: yellow purple
     Orientation of little circle in the triangle: up, right, left

Reinforcement

Get your student to create patterns with pattern blocks and/or tangrams.

Extra Practice, Unit 14, Exercise 2, pp. 193-194

Workbook

Exercise 6, pp. 181-182 (Answers p. 138)

In 2(d-e) the pattern involves a combination of the first and second making the third shape.

Test

Tests, Unit 14, Chapter 2, A and B, pp. 193-200

Workbook

Exercise 3, pp. 174-175

1. Answers can vary.

(b)

(c)

(d)

Exercise 4, pp. 176-178

1. (a) half circle, triangle
   (b) triangle, square
   (c) half circle, quarter circle
   (d) square, rectangle
   (e) quarter circle, rectangle

2. (b) answers can vary

(c)

(d)

(e)

Exercise 5, pp, 179-180

1.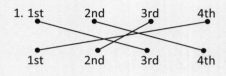

2.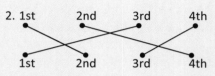

3. Answers will vary.

Exercise 6, pp. 181-182

1. (a)

(b)

(c)

(d)

2. (a)

(b)

(c)

(d)

(e)

Review 14

Review

Review 14, pp. 129-131

17: Set up the table for your student so that he just has to enter the tallies.

22: Let your student use counters or draw a picture.

24-25: If needed, show your student a rectangular prism and a pyramid.

Workbook

Review 15, pp. 183-192 (Answers p. 141)

1. (a) 825   (b) 300   (c) 541

2. (a) 160   (b) 159   (c) 303

3. (a) 16   (b) 70   (c) 28

4. (a) 8   (b) 6   (c) 10

5. (a) $10   (b) $1.09 (c) $1.50

6. (a) 18   (b) 15
   (c) 2   (d) 31
   (e) 72   (f) 83

7. $\dfrac{2}{5}$

8. (a) 8   (b) $\dfrac{1}{4}$

9. 8:20 p. m.

10. $1: 4 quarters
    $5: 4 x 5 = **20 quarters**

11. 340 g – 100 g = 240 g
    The box weighs **240 g**.

12. 275 – 206 = 69
    There were **69** children.

13. $8.25 - $6.50 = $1.75
    The shirt cost **$1.75** more.

14. 1 kg: $8
    5 kg: $8 x 5 = **$40**

15. (a) 3 chairs: $30
       1 chair: $30 ÷ 3 = **$10**.
    (b) $120 - $30 = $90
       The table cost **$90**.

16. (a) 168 + 287 = 455
       455 adults took part in the parade.
    (b) 455 – 113 = 342
       There were **342** more adults.

17.

winter	ⵜ⁄ I
spring	////
summer	ⵜ⁄ ⵜ⁄ ////
fall	ⵜ⁄

18. 302 qt + 29 qt = 331 qt
    331 quarts were sold this month.

19. 115 gal – 38 gal = 77 gal
    77 gallons more can be poured.

20. 4

21. (a) Check length of line drawn.
    (b) shorter

22. 28 ÷ 5 = 5 with 3 left over
    (28 – 5 – 5 – 5 – 5 – 5 = 3)
    She had **3** strawberries left.

23. 1 week: 7 days
    2 weeks: 7 x 2 = **14 days**

24. (a) 8
    (b) 12

25. (a) 4
    (b) 2 more

Review 15

Review

Review 15, pp. 132-138

19(b): The answer is greater than 1000; students have only had sums within 1000 before now, but your student may be able to extend addition of hundreds to the thousands place.

Test

Tests, Units 1-14, Cumulative, A and B, pp. 201-213

1. (a) 500   (b) 850   (c) 901

2. (a) 480   (b) 603   (c) 259

3. (a) 27   (b) 90   (c) 30

4. (a) 7   (b) 10   (c) 1

5. (a) $10
   (b) $9.20
   (c) $1.40

6. (a) 302   (b) 101
   (c) 566   (d) 225

7. (a) 70   (b) 200
   (c) 50   (d) 909

8. (a) 257, 275, 752
   (b) $\frac{1}{9}$, $\frac{1}{6}$, $\frac{1}{3}$

9. 4 fifty-cent coins = $2
   3 5-cent coins = $0.15
   The total money is **$2.15**.

10. 100 m − 48 m = 52 m
    He is **52 m** from the finishing point.

11. 10 hours

12. (a) 9
    (b) 14
    (c) 8

13. 1 dress: 2 hours
    5 dresses: 2 x 5 = **10 hours**

14. How many $2 cans in $16?
    16 ÷ 2 = 8
    She bought **8** cans.

15. $8.60 − $6.80 = $1.80
    She needs **$1.80**.

16. 305 − 46 = 259
    There were **259** girls.

17. 128 ℓ + 25 ℓ = 153 ℓ
    He used **153 liters** of gas.

18. 4 plates: $24
    1 plate: $24 ÷ 4 = **$6**

19. (a) $960 − $425 = $535
    The oven costs **$535**.
    (b) $960 + $535 = $1495
    The total cost is **$1495**.

20. (a) 1 day: 10 pages
    6 days: 10 x 6 = **60** pages
    (b) 60 + 24 = 84
    There were **84** pages.

21. (a) B and F
    (b) 36
    (c) 30
    (d) 6
    (e) 24

22. $\frac{3}{8}$

23. same as first triangle

24.

25. $\frac{1}{6}$

26. (a) 4   (b) $\frac{1}{4}$
    (c) 10   (d) $\frac{3}{10}$

27. 60 minutes

28. 4 quarts

29. no

30. (a) 14 oz − 3 oz = 11 oz
    The lettuce weighs **11 oz**.
    (b) 14 oz + 11 oz = 25 oz
    The two bags weigh **25 oz**.

31. 1 bag: 2 lb
    10 bags: 2 lb x 10 = **20 lb**

32. (a) 59 lb + 78 lb = 137 lb
    Her mother weighs **137 lb**.
    (b) 59 lb + 137 lb = 196 lb
    Together they weigh **196 lb**.

33. 28 qt ÷ 4 = 7 qt
    There are **7 quarts** in each.

34. 1 pudding: 2 c
    7 puddings: 2 c x 7 = 14 c
    He needs **14 cups** of milk.

35. 470 yd + 250 yd = 720 yd
    He walked **720 yards**.

36. (a) 65 lb + 127 lb + 88 lb = 280 lb
    Their total weight is **280 lb**.
    (b) 127 lb − 65 lb = 62 lb
    David is **62 lb** lighter.

37. (a) 50 ÷ 10 = 5
    She bought **5** boxes.
    (b) $9 x 5 = $45
    She paid **$45**.

Workbook

Review 15, pp. 183-192

1. first group: A, F, J, H
   second group: C, I, E
   third group: B, D, G

2. (a) 5
   (b) 3

3. Sections colored may vary.
   (a) (b) (c)

4. (a) 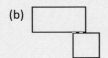 half-circle
      triangle

   (b) [rectangle figure] rectangle
      square

5. (a) 58   340   403   900
   (b) $\dfrac{1}{12}$ $\dfrac{1}{7}$ $\dfrac{1}{4}$ $\dfrac{1}{2}$

6. (a) 20   (b) 5
   (c) 30

7. [bar graph: Apple trees, Pear trees, Plum trees, Apricot trees, Peach trees; vertical axis 0 to 350]
   (a) 30
   (b) Apricot trees; 350
   (c) Peach trees; 80

8. (a) 45
   (b) 20
   (c) 60 + 45 + 40 + 50 + 45 = **240**

9. $\dfrac{3}{8}$

10. 405 − 240 = 165
    She made **165** more muffins.

11. 10 dolls: $70
    1 doll: $70 ÷ 10 = $7
    1 doll costs **$7**.

12. Group 20 apples by 4's.
    20 ÷ 4 = 5
    5 groups of 4 at $1 each.
    He paid **$5**.

13. 153 − 89 = 64
    She made **64** wheat cookies.

14. $10.40 − $3.95 = $6.45
    Her sister saved **$6.45**.

15. 1 bucket = 4 ℓ
    8 buckets = 4 ℓ x 8 = 32 ℓ
    The capacity is **32 liters**.

16. $8.20 − $6.90 = $1.30
    The radio was **$1.30** cheaper.

17. 26 ÷ 3 = 8 with 2 left over
    9 buckets are needed.

18. $5.20 − $1.80 = $3.40
    The notebook cost $**3.40**.

19. 5 bags: 40 lb
    1 bag: 40 lb ÷ 5 = **8 lb**

20. 17 gal + 25 gal = 42 gal
    The capacity of the tank is **42 gallons**.

21. $3.80 + $0.50 = $4.30
    The book costs **$4.30**.

22. 24 quarters = 24 ÷ 4 dollars = $6
    Amount of money Ryan has = $6.10.
    Yes, he has enough money.

23. Change = 50¢ − 29¢ = **21¢**

24. 32 quarters = 32 ÷ 4 = 8 dollars
    $8 + $1.20 + $0.30 + $0.03 = $9.53
    She has $**9.53**.

Potentially challenging problems in the test book, first printing 2008.

Page	Test	Number	Notes
11	Unit 7 Ch. 3, Test B	3	482 – 199. Students have learned to subtract 99, not 199, but your student may be able to extend the concepts to this situation (which will be taught in *Primary Mathematics* 3).
41	Units 1-8, Test A	6(a, c)	This is potentially a 2-step problem. The answer to 6(a) must be correct to have a correct answer to 6(c).
49	Unit 9, Ch. 1, Test B	3	This involves adding dollars to dollars and cents, which is specifically taught in the next chapter, but will probably not pose any difficulty.
100	Units 1-10, Test B	17	Allow your student to use counters or a drawing with this one, if needed. This is a 2-step problem and could be quite challenging at this level.
137	Units 1-11, Test B	9	This problem may be challenging since your student cannot look at any single balance to get the answer. It may look like A is heaviest, since the balance with it goes down the farthest, but the middle balance shows that C is heavier than A.
146	Unit 12, Ch. 1, Test B	8	This is a 2-step problem.
162	Units 1-12, Test A	13	Your student will have to know the number of days in each month.
163	Units 1-12, Test A	14	This is a 2-step problem.
188	Unit 14, Ch. 1, Test A	2(b)	Insert the word "flat" before faces, if it is not there. A _____ has no edges and no *flat* faces.
205	Units 1-14, Test A	15	Might be challenging.
205	Units 1-14, Test A	16	Your student has to be careful to actually read the problem, not rush into an answer from the picture.
206	Units 1-14, Test A	19	This is a 2-step problem.
211	Units 1-14, Test B	14	Allow your student to use counters if needed.

Mental Math 1	Mental Math 2	Mental Math 3		Mental Math 4	Mental Math 5	Mental Math 6
100 − 25 = **75**	37 + **63** = 100	100 − 11 = **89**		24 + 7 = **31**	18 + 80 = **98**	432 + 4 = **436**
100 − 50 = **50**	76 + **24** = 100	5 + **8** = 13		72 + 5 = **77**	42 + 20 = **62**	216 + 4 = **220**
100 − 75 = **25**	51 + **49** = 100	100 − **2** = 98		65 + 6 = **71**	68 + 20 = **88**	724 + 6 = **730**
100 − 99 = **1**	40 + **60** = 100	**62** + 38 = 100		59 + 5 = **64**	72 + 10 = **82**	908 + 4 = **912**
100 − 5 = **95**	**10** + 90 = 100	84 + **16** = 100		66 + 7 = **73**	88 + 20 = **108**	112 + 4 = **116**
100 − 60 = **40**	**45** + 55 = 100	4 + **14** = 18		25 + 3 = **28**	12 + 80 = **92**	309 + 1 = **310**
100 − 35 = **65**	**93** + 7 = 100	25 + **75** = 100		64 + 9 = **73**	54 + 90 = **144**	504 + 9 = **513**
100 − 45 = **55**	**67** + 33 = 100	100 − **30** = 70		77 + 8 = **85**	86 + 20 = **106**	758 + 2 = **760**
100 − 44 = **56**	100 − **20** = 80	8 + **4** = 12		71 + 9 = **80**	32 + 50 = **82**	849 + 4 = **853**
100 − 10 = **90**	100 − **59** = 41	**50** + 50 = 100		35 + 8 = **43**	52 + 60 = **112**	285 + 6 = **291**
100 − 87 = **13**	100 − **18** = 82	100 − 15 = **85**		57 + 5 = **62**	77 + 70 = **147**	847 + 7 = **854**
100 − 22 = **78**	100 − **43** = 57	**12** − 5 = 7		35 + 5 = **40**	82 + 40 = **122**	258 + 4 = **262**
100 − 4 = **96**	100 − **82** = 18	100 − **72** = 28		56 + 9 = **65**	60 + 60 = **120**	185 + 7 = **192**
100 − 69 = **31**	100 − **29** = 71	6 + **9** = 15		42 + 6 = **48**	78 + 90 = **168**	987 + 8 = **995**
100 − 95 = **5**	85 + **15** = 100	**54** + 46 = 100		86 + 6 = **92**	87 + 50 = **137**	618 + 5 = **623**
100 − 70 = **30**	87 + **13** = 100	100 − **25** = 75		59 + 7 = **66**	60 + 72 = **132**	526 + 7 = **533**
100 − 23 = **77**	**65** + 35 = 100	75 + 25 + 94 = **194**		37 + 3 = **40**	70 + 87 = **157**	857 + 9 = **866**
100 − 7 = **93**	8 + **92** = 100	69 + 27 + 31 = **127**		79 + 3 = **82**	85 + 30 = **115**	146 + 4 = **150**
100 − 66 = **34**	100 − **97** = 3	54 + 46 + **19** = 119		28 + 6 = **34**	56 + 90 = **146**	493 + 4 = **497**
100 − 31 = **69**	100 − **29** = 71	57 + 72 + **28** = 157		37 + 7 = **44**	74 + 40 = **114**	307 + 4 = **311**

Mental Math 7	Mental Math 8	Mental Math 9		Mental Math 10	Mental Math 11	Mental Math 12
105 + 500 = **605**	24 + 72 = **96**	24 + 99 = **123**		99 + 481 = **580**	90 − 9 = **81**	99 − 20 = **79**
136 + 400 = **536**	72 + 15 = **87**	25 + 98 = **123**		324 + 99 = **423**	30 − 8 = **22**	34 − 10 = **24**
758 + 200 = **958**	65 + 32 = **97**	5 + 98 = **103**		182 + 98 = **280**	20 − 7 = **13**	82 − 4 = **78**
412 + 300 = **712**	53 + 45 = **98**	32 + 99 = **131**		99 + 304 = **403**	70 − 4 = **66**	75 − 3 = **72**
111 + 700 = **811**	66 + 32 = **98**	67 + 99 = **166**		126 + 98 = **224**	80 − 5 = **75**	41 − 6 = **35**
626 + 300 = **926**	25 + 62 = **87**	25 + 97 = **122**		99 + 412 = **511**	20 − 2 = **18**	84 − 9 = **75**
614 + 80 = **694**	64 + 12 = **76**	42 + 98 = **140**		98 + 626 = **724**	60 − 7 = **53**	31 − 7 = **24**
248 + 20 = **268**	71 + 28 = **99**	71 + 98 = **169**		197 + 98 = **295**	40 − 3 = **37**	97 − 9 = **88**
160 + 60 = **220**	17 + 82 = **99**	62 + 99 = **161**		328 + 99 = **427**	50 − 8 = **42**	66 − 8 = **58**
972 + 10 = **982**	35 + 63 = **98**	45 + 98 = **143**		845 + 98 = **943**	90 − 6 = **84**	73 − 8 = **65**
622 + 70 = **692**	57 + 41 = **98**	97 + 99 = **196**		107 + 99 = **206**	40 − 5 = **35**	52 − 5 = **47**
140 + 60 = **200**	16 + 73 = **89**	75 + 99 = **174**		375 + 95 = **470**	30 − 1 = **29**	54 − 7 = **47**
590 + 40 = **630**	41 + 26 = **67**	16 + 97 = **113**		210 + 94 = **304**	50 − 5 = **45**	23 − 9 = **14**
269 + 20 = **289**	82 + 11 = **93**	41 + 98 = **139**		648 + 92 = **740**	70 − 9 = **61**	85 − 8 = **77**
370 + 50 = **420**	81 + 16 = **97**	12 + 99 = **111**		820 + 93 = **913**	47 − 6 = **41**	42 − 6 = **36**
452 + 60 = **512**	52 + 24 = **76**	99 + 16 = **115**		98 + 691 = **789**	89 − 4 = **85**	67 − 3 = **64**
778 + 70 = **848**	37 + 31 = **68**	97 + 34 = **131**		98 + 231 = **329**	96 − 5 = **91**	35 − 6 = **29**
285 + 40 = **325**	27 + 62 = **89**	67 + 96 = **163**		368 + 96 = **464**	63 − 30 = **33**	72 − 9 = **63**
364 + 60 = **424**	35 + 5 = **40**	95 + 38 = **133**		95 + 238 = **333**	97 − 50 = **47**	47 − 8 = **39**
578 + 90 = **668**	32 + 28 = **60**	99 + 2 = **101**		96 + 602 = **698**	86 − 70 = **16**	41 − 2 = **39**

Answers to Mental Math

Mental Math 13	Mental Math 14	Mental Math 15
290 – 9 = **281**	500 – 60 = **440**	96 – 72 = **24**
530 – 8 = **522**	200 – 70 = **130**	24 – 17 = **7**
120 – 7 = **113**	400 – 20 = **380**	86 – 29 = **57**
970 – 9 = **961**	700 – 60 = **640**	78 – 36 = **42**
820 – 5 = **815**	500 – 70 = **430**	51 – 38 = **13**
426 – 7 = **419**	520 – 60 = **460**	47 – 26 = **21**
653 – 7 = **646**	220 – 70 = **150**	74 – 52 = **22**
364 – 6 = **358**	410 – 20 = **390**	89 – 54 = **35**
752 – 8 = **744**	750 – 60 = **690**	93 – 61 = **32**
373 – 6 = **367**	532 – 70 = **462**	85 – 55 = **30**
249 – 5 = **244**	220 – 80 = **140**	42 – 16 = **26**
481 – 8 = **473**	840 – 20 = **820**	35 – 11 = **24**
556 – 5 = **551**	621 – 20 = **601**	73 – 66 = **7**
176 – 9 = **167**	348 – 60 = **288**	47 – 19 = **28**
668 – 6 = **662**	732 – 90 = **642**	94 – 32 = **62**
884 – 9 = **875**	720 – 30 = **690**	84 – 59 = **25**
391 – 7 = **384**	930 – 20 = **910**	91 – 74 = **17**
927 – 9 = **918**	423 – 400 = **23**	82 – 58 = **24**
646 – 8 = **638**	632 – 400 = **232**	96 – 35 = **61**
713 – 8 = **705**	716 – 200 = **516**	82 – 45 = **37**

Mental Math 16	Mental Math 17	Mental Math 18
200 – 99 = **101**	431 – 98 = **333**	32 – 5 = **27**
400 – 98 = **302**	698 – 98 = **600**	120 + 80 = **200**
600 – 99 = **501**	843 – 99 = **744**	88 – 30 = **58**
500 – 99 = **401**	259 – 98 = **161**	34 + 61 = **95**
100 – 98 = **2**	742 – 98 = **644**	960 + 40 = **1000**
800 – 99 = **701**	365 – 97 = **268**	100 – 77 = **23**
300 – 97 = **203**	785 – 98 = **687**	340 + 90 = **430**
700 – 98 = **602**	256 – 99 = **157**	289 + 400 = **689**
900 – 97 = **803**	568 – 98 = **470**	284 + 98 = **382**
600 – 96 = **504**	762 – 97 = **665**	306 – 9 = **297**
400 – 97 = **303**	369 – 97 = **272**	362 + 7 = **369**
600 – 95 = **505**	258 – 99 = **159**	411 + 99 = **510**
803 – 99 = **704**	204 – 98 = **106**	632 – 99 = **533**
206 – 98 = **108**	560 – 96 = **464**	640 – 80 = **560**
409 – 98 = **311**	701 – 9	
305 – 99 = **206**	584 – 98	
705 – 97 = **608**	887 – 95	
208 – 97 = **111**	963 – 93	
108 – 98 = **10**	514 – 94	
702 – 96 = **606**	681 – 95	

Mental Math 19	Mental Math 20
58 + 7 = **65**	834 + 7 – 20 + 60 = **881**
85 – 32 = **53**	368 – 99 + 40 + 3 = **312**
235 + 70 = **305**	429 + 8 – 7 – 90 = **340**
984 – 200 = **784**	501 – 4 – 30 + 99 = **566**
984 – 121 = **863**	87 + 7 + 2 – 9 + 40 = **127**
587 – 98 = **489**	4 + 4 + 4 + 4 + 98 = **114**
203 + 8 = **211**	5 + 5 + 5 + 5 + 280 = **300**
698 – 3 = **695**	33 + 21 + 9 – 30 + 130 = **163**
100 – 28 = **72**	925 – 98 + 3 – 70 + 5 = **765**
98 + 654 = **752**	62 + 8 – 2 + 40 – 8 – 33 = **67**
687 – 99 = **588**	592 + 99 – 98 + 7 = **600**
414 + 8 = **422**	83 + 7 – 26 + 99 = **163**
683 – 7 = **676**	300 – 99 + 70 + 6 = **277**
610 – 90 = **520**	29 + 38 – 17 + 20 = **70**
700 + 98 = **798**	203 – 7 – 3 + 400 = **593**
250 – 98 = **152**	3 + 3 + 3 + 3 + 3 + 2 = **17**
250 + 80 = **330**	4 + 4 + 4 + 4 + 4 + 4 = **24**
100 – 62 = **38**	5 + 5 + 5 + 5 + 5 + 5 = **30**
754 – 601 = **153**	40 – 4 – 4 – 4 – 4 – 4 – 4 = **16**
986 – 123 = **863**	2 + 3 + 4 + 5 + 6 + 7 + 8 + 9 + 10 = **54**

Mental Math 21	Mental Math 22	Mental Math 23
2 x 4 = **8**	2 x 7 = **14**	12 ÷ 4 = **3**
4 x 9 = **36**	4 x 8 = **32**	4 ÷ 4 = **1**
6 x 4 = **24**	9 x 3 = **27**	36 ÷ 4 = **9**
4 x 7 = **28**	4 x 5 = **20**	32 ÷ 4 = **8**
4 x 6 = **24**	2 x 6 = **12**	24 ÷ 4 = **6**
4 x 3 = **12**	8 x 2 = **16**	40 ÷ 4 = **10**
7 x 4 = **28**	4 x 6 = **24**	20 ÷ 4 = **5**
4 x 10 = **40**	9 x 4 = **36**	4 ÷ 4 = **1**
4 x 5 = **20**	5 x 3 = **15**	16 ÷ 4 = **4**
8 x 4 = **32**	2 x 9 = **18**	8 ÷ 4 = **2**
5 x 4 = **20**	4 x 4 = **16**	16 ÷ 4 = **4**
4 x 9 = **36**	3 x 8 = **24**	28 ÷ 4 = **7**
4 x 4 = **16**	6 x 3 = **18**	32 ÷ 4 = **8**
7 x 4 = **28**	2 x 5 = **10**	8 ÷ 4 = **2**
4 x 8 = **32**	3 x 2 = **6**	12 ÷ 4 = **3**
4 x 6 = **24**	3 x 7 = **21**	20 ÷ 4 = **5**
4 x 10 = **40**	4 x 3 = **12**	28 ÷ 4 = **7**
4 x 2 = **8**	3 x 3 = **9**	16 ÷ 4 = **4**
8 x 4 = **32**	7 x 4 = **28**	24 ÷ 4 = **6**
3 x 4 = **12**	2 x 4 = **8**	40 ÷ 4 = **10**

Mental Math 24	Mental Math 25	Mental Math 26		Mental Math 27	Mental Math 28	Mental Math 29
12 ÷ 3 = **4**	4 x 4 = **16**	1 x 5 = **5**		3 x 6 = **18**	35 ÷ 5 = **7**	30 ÷ 5 = **6**
36 ÷ 4 = **9**	36 ÷ 4 = **9**	6 x 5 = **30**		5 x 2 = **10**	27 ÷ 3 = **9**	35 ÷ 5 = **7**
18 ÷ 2 = **9**	5 x 4 = **20**	4 x 5 = **20**		7 x 4 = **28**	5 ÷ 5 = **1**	10 ÷ 2 = **5**
24 ÷ 4 = **6**	28 ÷ 4 = **7**	5 x 7 = **35**		3 x 5 = **15**	16 ÷ 2 = **8**	40 ÷ 5 = **8**
15 ÷ 3 = **5**	30 ÷ 3 = **10**	5 x 2 = **10**		3 x 9 = **27**	15 ÷ 5 = **3**	32 ÷ 4 = **8**
32 ÷ 4 = **8**	4 x 7 = **28**	3 x 5 = **15**		5 x 4 = **20**	36 ÷ 4 = **9**	50 ÷ 5 = **10**
18 ÷ 3 = **6**	9 x 4 = **36**	5 x 10 = **50**		3 x 4 = **12**	25 ÷ 5 = **5**	20 ÷ 2 = **10**
16 ÷ 2 = **8**	24 ÷ 4 = **6**	8 x 5 = **40**		5 x 5 = **25**	9 ÷ 3 = **3**	25 ÷ 5 = **5**
40 ÷ 4 = **10**	4 x 6 = **24**	5 x 5 = **25**		4 x 4 = **16**	30 ÷ 5 = **6**	10 ÷ 5 = **2**
20 ÷ 4 = **5**	20 ÷ 4 = **5**	10 x 5 = **50**		6 x 5 = **30**	18 ÷ 2 = **9**	15 ÷ 3 = **5**
21 ÷ 3 = **7**	3 x 8 = **24**	5 x 3 = **15**		5 x 7 = **35**	45 ÷ 5 = **9**	45 ÷ 5 = **9**
28 ÷ 4 = **7**	18 ÷ 3 = **6**	5 x 6 = **30**		4 x 8 = **32**	24 ÷ 3 = **8**	20 ÷ 5 = **4**
24 ÷ 3 = **8**	21 ÷ 3 = **7**	5 x 9 = **45**		9 x 5 = **45**	50 ÷ 5 = **10**	24 ÷ 3 = **8**
27 ÷ 3 = **9**	4 x 5 = **20**	2 x 5 = **10**		4 x 9 = **36**	16 ÷ 4 = **4**	40 ÷ 4 = **10**
24 ÷ 4 = **6**	8 ÷ 4 = **2**	5 x 5 = **25**		10 x 5 = **50**	30 ÷ 3 = **10**	15 ÷ 5 = **3**
8 ÷ 4 = **2**	32 ÷ 4 = **8**	7 x 5 = **35**		7 x 3 = **21**	12 ÷ 2 = **6**	18 ÷ 3 = **6**
16 ÷ 4 = **4**	10 x 4 = **40**	5 x 1 = **5**		4 x 2 = **8**	40 ÷ 5 = **8**	5 ÷ 5 = **1**
20 ÷ 2 = **10**	12 ÷ 4 = **3**	5 x 8 = **40**		5 x 8 = **40**	10 ÷ 5 = **2**	12 ÷ 3 = **4**
16 ÷ 4 = **4**	40 ÷ 4 = **10**	9 x 5 = **45**		6 x 4 = **24**	28 ÷ 4 = **7**	21 ÷ 3 = **7**
12 ÷ 2 = **6**	7 x 3 = **21**	5 x 4 = **20**		10 x 4 = **40**	20 ÷ 5 = **4**	12 ÷ 4 = **3**

Mental Math 30	Mental Math 31	Mental Math 32		Mental Math 33	Mental Math 34	Mental Math 35
10 x 7 = **70**	20 ÷ 10 = **2**	15 ÷ 5 = **3**		2 x 3 = **6**	100 − 29 = **71**	100 − 75 = **25**
5 x 8 = **40**	100 ÷ 10 = **10**	28 ÷ 4 = **7**		12 ÷ 4 = **3**	876 + 7 = **883**	50 ÷ 5 = **10**
10 x 2 = **20**	30 ÷ 3 = **10**	5 x 9 = **45**		80 ÷ 10 = **8**	24 ÷ 4 = **6**	73 + 8 = **81**
3 x 9 = **27**	30 ÷ 10 = **3**	4 x 5 = **20**		8 ÷ 2 = **4**	68 + 90 = **158**	68 + 80 = **148**
4 x 10 = **40**	10 ÷ 5 = **2**	20 ÷ 4 = **5**		2 x 5 = **10**	361 + 50 = **411**	216 + 9 = **225**
5 x 10 = **50**	50 ÷ 10 = **5**	32 ÷ 4 = **8**		100 ÷ 10 = **10**	637 + 41 = **678**	167 + 70 = **237**
8 x 4 = **32**	10 ÷ 2 = **5**	4 x 4 = **16**		24 ÷ 3 = **8**	5 x 9 = **45**	37 + 31 = **68**
10 x 6 = **60**	70 ÷ 10 = **7**	25 ÷ 5 = **5**		21 ÷ 3 = **7**	98 + 238 = **336**	23 − 9 = **14**
9 x 5 = **45**	30 ÷ 5 = **6**	6 x 5 = **30**		24 ÷ 4 = **6**	42 − 6 = **36**	420 − 8 = **412**
8 x 10 = **80**	60 ÷ 10 = **6**	4 x 2 = **8**		3 x 3 = **9**	954 − 7 = **947**	920 − 50 = **870**
7 x 2 = **14**	50 ÷ 5 = **10**	80 ÷ 10 = **8**		3 x 6 = **18**	123 − 90 = **33**	10 x 9 = **90**
10 x 10 = **100**	10 ÷ 10 = **1**	4 x 6 = **24**		15 ÷ 3 = **5**	47 − 16 = **31**	57 − 35 = **22**
9 x 4 = **36**	40 ÷ 5 = **8**	10 x 7 = **70**		8 x 10 = **80**	762 − 98 = **664**	584 − 98 = **486**
8 x 3 = **24**	40 ÷ 4 = **10**	20 ÷ 5 = **4**		27 ÷ 3 = **9**	800 − 99 = **701**	986 − 400 = **586**
10 x 1 = **10**	90 ÷ 10 = **9**	10 x 6 = **60**		32 ÷ 4 = **8**	5 x 4 = **20**	95 + 198 = **293**
7 x 3 = **21**	20 ÷ 5 = **4**	4 x 8 = **32**		4 x 3 = **12**	35 ÷ 5 = **7**	36 ÷ 4 = **9**
3 x 10 = **30**	40 ÷ 10 = **4**	10 x 5 = **50**		24 ÷ 3 = **8**	59 + 5 = **64**	7 x 5 = **35**
9 x 10 = **90**	20 ÷ 4 = **5**	8 x 5 = **40**		4 x 7 = **28**	7 x 4 = **28**	34 + 61 = **95**
4 x 7 = **28**	80 ÷ 10 = **8**	24 ÷ 3 = **8**		10 x 10 = **100**	100 ÷ 10 = **10**	232 + 600 = **832**
5 x 7 = **35**	20 ÷ 2 = **10**	3 x 5 = **15**		6 x 10 = **60**	683 − 7 = **676**	4 x 8 = **32**

Mental Math 36	Mental Math 37		Mental Math 38	Mental Math 39
75¢ = **3** quarters	$1 – 75¢ = **25** ¢		$10 – $0.85 = $ **9.15**	$2.65 + $3 = $ **5.65**
60¢ = **6** dimes	$1 – 50¢ = **50** ¢		$10 – $0.40 = $ **9.60**	$1.25 + 75¢ = $ **2.00**
55¢ = **11** nickels	$1 – 35¢ = **65** ¢		$10 – $2.30 = $ **7.70**	$3.15 + 55¢ = $ **3.70**
50¢ = **2** quarters	$1 – 60¢ = **40** ¢		$3.60 + $ **6.40** = $10	$5.32 + $2 = $ **7.32**
$1 = **4** quarters	$1 – 25¢ = **75** ¢		$8.35 + $ **1.65** = $10	$8.35 + 25¢ = $ **8.60**
$6 = **24** quarters	85¢ + **15** ¢= $1		$2.70 + $ **7.30** = $10	$1.32 + 27¢ = $ **1.59**
$2 = **40** nickels	45¢ + **55** ¢ = $1		$10 – $3.65 = $ **6.35**	$2.65 + $4 = $ **6.65**
$3 = **300** pennies	10¢ + **90** ¢ = $1		$10 – $4.20 = $ **5.80**	$2.41 + 32¢ = $ **2.73**
$5 = **50** dimes	**70** ¢ + 30¢ = $1		$10 – $2.05 = $ **7.95**	$9.20 + 62¢ = $ **9.82**
$3.80 = **38** dimes	11¢ + **89** ¢ = $1		$7.10 + $ **2.90** = $10	$3.20 + $4.40 = $ **7.60**
$4.25 = **17** quarters	**93** ¢ + 7¢ = $1		$8.35 + $ **1.65** = $10	$3.35 + $2.15 = $ **5.50**
$1.20 = **24** nickels	$1 – 32¢ = **68** ¢		$ **5.85** + $4.15 = $10	$6.40 + $0.15 = $ **6.55**
$8.50 = **34** quarters	49¢ + **51** ¢ = $1		$10 – $5.75 = $ **4.25**	$2.25 + $4.35 = $ **6.60**
$12.50 = **125** dimes	94¢ + **6** ¢ = $1		$4.93 + $ **5.07** = $10	$2.40 + $3.18 = $ **5.58**
$20 = **4** 5-dollar bills	$1 – 56¢ = **44** ¢		$9.36 + $ **0.64** = $10	$6.24 + $1.07 = $ **7.31**
$10 = **10** 1-dollar bills	**28** ¢ + 72¢ = $1		$10 – $5.61 = $ **4.39**	$2.41 + $6.25 = $ **8.66**
$40 = **2** 20-dollar bills	$1 – 13¢ = **87** ¢		$ **2.72** + $7.28 = $10	$4.35 + $4.49 = $ **8.84**
$100 = **10** 10-dollar bills	**36** ¢ + 64¢ = $1		$10 – $0.12 = $ **9.88**	$2.05 + $6.95 = $ **9.00**
$45 = **9** 5-dollar bills	**72** ¢ + 28¢ = $1		$ **1.56** + $8.44 = $10	$0.30 + $6.70 = $ **7.00**
$30 = **6** 5-dollar bills			$ **8.73** + $1.27 = $10	$2.45 + $5.55 = $ **8.00**

Mental Math 40	Mental Math 41		Mental Math 42	Mental Math 43
$6.40 + $0.99 = $ **7.39**	$8.60 – $4 = $ **4.60**		$3 – $0.99 = $ **2.01**	$6.35 – $1.97 = $ **4.38**
$2.25 + $0.98 = $ **3.23**	68¢ – 21¢ = $ **0.47**		$5 – $0.96 = $ **4.04**	$348 + $98 = $ **446**
$2.43 + $3.98 = $ **6.41**	$9.35 – $7 = $ **2.35**		$8 – $0.97 = $ **7.03**	$329 – $98 = $ **231**
$6.24 + $1.97 = $ **8.21**	$4 – 75¢ = $ **3.25**		$4.35 – $0.98 = $ **3.37**	$3.05 + $6.95 = $ **10.00**
$2.20 + $6.95 = $ **9.15**	$8 – 25¢ = $ **7.75**		$7.25 – $0.96 = $ **6.29**	$10.00 – $6.40 = $ **3.60**
$4.35 + $4.99 = $ **9.34**	$5.15 – $3 = $ **2.15**		$3.65 – $0.95 = $ **2.70**	$4.65 + $3.99 = $ **8.64**
$2.05 + $6.95 = $ **9.00**	$5 – 40¢ = $ **4.60**		$5.80 – $3.96 = $ **1.84**	$1.00 – $0.35 = $ **0.65**
$3.45 + $2.96 = $ **6.41**	$3.83 – 25¢ = $ **3.58**		$9.05 – $6.97 = $ **2.08**	$1.00 – $0.85 = $ **0.15**
$1.25 + $0.97 = $ **2.22**	$3 – 29¢ = $ **2.71**		$6.70 – $2.95 = $ **3.75**	$10.00 – $2.75 =$ **7.25**
$4.65 + $3.98 = $ **8.63**	$4.90 – $1.80 = $ **3.10**		$2.85 – $0.97 = $ **1.88**	$9.27 + $0.66 = $ **9.93**
$6.97 + $1.24 = $ **8.21**	$8.85 – $6.05 = $ **2.80**		$7.69 – $3.95 = $ **3.74**	$3.83 – $2.25 = $ **1.58**
$7.30 + $0.95 = $ **8.25**	$3.45 – $1.25 = $ **2.20**		$6.25 – $0.95 = $ **5.30**	$3.21 + $4.72 = $ **7.93**
$2.96 + $1.37 = $ **4.33**	$9.95 – $4.30 = $ **5.65**		$4.85 – $3.96 = $ **0.89**	$3.35 + $2.99 = $ **6.34**
$2.05 + $6.95 = $ **9.00**	$7.70 – $3.35 = $ **4.35**		$9.32 – $6.95 = $ **2.37**	$5.00 – $0.45 = $ **4.55**
$1.07 + $3.97 = $ **5.04**	$6.65 – $2.45 = $ **4.20**		$6.29 – $2.95 = $ **3.34**	$3.00 – $0.35 = $ **2.65**
$3.96 + $2.55 = $ **6.51**	$3.85 – $2.15 = $ **1.70**		$5.82 – $0.97 = $ **4.85**	$4.05 – $1.75 = $ **2.30**
$4.95 + $2.24 = $ **7.19**	$42 – $8 = $ **34**		$6.67 – $5.95 = $ **0.72**	$2.95 + $3.50 = $ **6.45**
$0.20 + $1.96 = $ **2.16**	$95 – $7 = $ **88**		$8.35 – $2.96 = $ **5.39**	$8.75 – $6.50 = $ **2.25**
$2.95 + $3.37 = $ **6.32**	$387 – $98 = $ **289**		$1.98 – $0.99 = $ **0.99**	$1.25 + $0.90 = $ **2.15**
$1.95 + $6.99 = $ **8.94**	$600 – $42 = $ **558**		$9.55 – $8.95 = $ **0.60**	$3.65 + $5.75 = $ **9.40**

Mental Math 1	Mental Math 2	Mental Math 3
100 – 25 = _____	37 + _____ = 100	100 – 11 = _____
100 – 50 = _____	76 + _____ = 100	5 + _____ = 13
100 – 75 = _____	51 + _____ = 100	100 – _____ = 98
100 – 99 = _____	40 + _____ = 100	_____ + 38 = 100
100 – 5 = _____	_____ + 90 = 100	84 + _____ = 100
100 – 60 = _____	_____ + 55 = 100	4 + _____ = 18
100 – 35 = _____	_____ + 7 = 100	25 + _____ = 100
100 – 45 = _____	_____ + 33 = 100	100 – 30 = _____
100 – 44 = _____	100 – _____ = 80	8 + _____ = 12
100 – 10 = _____	100 – _____ = 41	_____ + 50 = 100
100 – 87 = _____	100 – _____ = 82	100 – 15 = _____
100 – 22 = _____	100 – _____ = 57	_____ – 5 = 7
100 – 4 = _____	100 – _____ = 18	100 – 72 = _____
100 – 69 = _____	100 – _____ = 71	6 + _____ = 15
100 – 95 = _____	85 + _____ = 100	_____ + 46 = 100
100 – 70 = _____	_____ + 13 = 100	100 – _____ = 75
100 – 23 = _____	_____ + 35 = 100	75 + 25 + 94 = _____
100 – 7 = _____	8 + _____ = 100	69 + 27 + 31 = _____
100 – 66 = _____	100 – _____ = 3	54 + 46 + _____ = 119
100 – 31 = _____	100 – 29 = _____	57 + 72 + _____ = 157

Mental Math 4	Mental Math 5	Mental Math 6
24 + 7 = _____	18 + 80 = _____	432 + 4 = _____
72 + 5 = _____	42 + 20 = _____	216 + 4 = _____
65 + 6 = _____	68 + 20 = _____	724 + 6 = _____
59 + 5 = _____	72 + 10 = _____	908 + 4 = _____
66 + 7 = _____	88 + 20 = _____	112 + 4 = _____
25 + 3 = _____	12 + 80 = _____	309 + 1 = _____
64 + 9 = _____	54 + 90 = _____	504 + 9 = _____
77 + 8 = _____	86 + 20 = _____	758 + 2 = _____
71 + 9 = _____	32 + 50 = _____	849 + 4 = _____
35 + 8 = _____	52 + 60 = _____	285 + 6 = _____
57 + 5 = _____	77 + 70 = _____	847 + 7 = _____
35 + 5 = _____	82 + 40 = _____	258 + 4 = _____
56 + 9 = _____	60 + 60 = _____	185 + 7 = _____
42 + 6 = _____	78 + 90 = _____	987 + 8 = _____
86 + 6 = _____	87 + 50 = _____	618 + 5 = _____
59 + 7 = _____	60 + 72 = _____	526 + 7 = _____
37 + 3 = _____	70 + 87 = _____	857 + 9 = _____
79 + 3 = _____	85 + 30 = _____	146 + 4 = _____
28 + 6 = _____	56 + 90 = _____	493 + 4 = _____
37 + 7 = _____	74 + 40 = _____	307 + 4 = _____

Mental Math 7	Mental Math 8	Mental Math 9
105 + 500 = _____	24 + 72 = _____	24 + 99 = _____
136 + 400 = _____	72 + 15 = _____	25 + 98 = _____
758 + 200 = _____	65 + 32 = _____	5 + 98 = _____
412 + 300 = _____	53 + 45 = _____	32 + 99 = _____
111 + 700 = _____	66 + 32 = _____	67 + 99 = _____
626 + 300 = _____	25 + 62 = _____	25 + 97 = _____
614 + 80 = _____	64 + 12 = _____	42 + 98 = _____
248 + 20 = _____	71 + 28 = _____	71 + 98 = _____
160 + 60 = _____	17 + 82 = _____	62 + 99 = _____
972 + 10 = _____	35 + 63 = _____	45 + 98 = _____
622 + 70 = _____	57 + 41 = _____	97 + 99 = _____
140 + 60 = _____	16 + 73 = _____	75 + 99 = _____
590 + 40 = _____	41 + 26 = _____	16 + 97 = _____
269 + 20 = _____	82 + 11 = _____	41 + 98 = _____
370 + 50 = _____	81 + 16 = _____	12 + 99 = _____
452 + 60 = _____	52 + 24 = _____	99 + 16 = _____
778 + 70 = _____	37 + 31 = _____	97 + 34 = _____
285 + 40 = _____	27 + 62 = _____	67 + 96 = _____
364 + 60 = _____	35 + 5 = _____	95 + 38 = _____
578 + 90 = _____	32 + 28 = _____	99 + 2 = _____

Mental Math 10	Mental Math 11	Mental Math 12
99 + 481 = _____	90 − 9 = _____	99 − 20 = _____
324 + 99 = _____	30 − 8 = _____	34 − 10 = _____
182 + 98 = _____	20 − 7 = _____	82 − 4 = _____
99 + 304 = _____	70 − 4 = _____	75 − 3 = _____
126 + 98 = _____	80 − 5 = _____	41 − 6 = _____
99 + 412 = _____	20 − 2 = _____	84 − 9 = _____
98 + 626 = _____	60 − 7 = _____	31 − 7 = _____
197 + 98 = _____	40 − 3 = _____	97 − 9 = _____
328 + 99 = _____	50 − 8 = _____	66 − 8 = _____
845 + 98 = _____	90 − 6 = _____	73 − 8 = _____
107 + 99 = _____	40 − 5 = _____	52 − 5 = _____
375 + 95 = _____	30 − 1 = _____	54 − 7 = _____
210 + 94 = _____	50 − 5 = _____	23 − 9 = _____
648 + 92 = _____	70 − 9 = _____	85 − 8 = _____
820 + 93 = _____	47 − 6 = _____	42 − 6 = _____
98 + 691 = _____	89 − 4 = _____	67 − 3 = _____
98 + 231 = _____	96 − 5 = _____	35 − 6 = _____
368 + 96 = _____	63 − 30 = _____	72 − 9 = _____
95 + 238 = _____	97 − 50 = _____	47 − 8 = _____
96 + 602 = _____	86 − 70 = _____	41 − 2 = _____

Mental Math 13	Mental Math 14	Mental Math 15
290 – 9 = _____	500 – 60 = _____	96 – 72 = _____
530 – 8 = _____	200 – 70 = _____	24 – 17 = _____
120 – 7 = _____	400 – 20 = _____	86 – 29 = _____
970 – 9 = _____	700 – 60 = _____	78 – 36 = _____
820 – 5 = _____	500 – 70 = _____	51 – 38 = _____
426 – 7 = _____	520 – 60 = _____	47 – 26 = _____
653 – 7 = _____	220 – 70 = _____	74 – 52 = _____
364 – 6 = _____	410 – 20 = _____	89 – 54 = _____
752 – 8 = _____	750 – 60 = _____	93 – 61 = _____
373 – 6 = _____	532 – 70 = _____	85 – 55 = _____
249 – 5 = _____	220 – 80 = _____	42 – 16 = _____
481 – 8 = _____	840 – 20 = _____	35 – 11 = _____
556 – 5 = _____	621 – 20 = _____	73 – 66 = _____
176 – 9 = _____	348 – 60 = _____	47 – 19 = _____
668 – 6 = _____	732 – 90 = _____	94 – 32 = _____
884 – 9 = _____	720 – 30 = _____	84 – 59 = _____
391 – 7 = _____	930 – 20 = _____	91 – 74 = _____
927 – 9 = _____	423 – 400 = _____	82 – 58 = _____
646 – 8 = _____	632 – 400 = _____	96 – 35 = _____
713 – 8 = _____	716 – 200 = _____	82 – 45 = _____

Mental Math 16	Mental Math 17	Mental Math 18
200 – 99 = _____	431 – 98 = _____	32 – 5 = _____
400 – 98 = _____	698 – 98 = _____	120 + 80 = _____
600 – 99 = _____	843 – 99 = _____	88 – 30 = _____
500 – 99 = _____	259 – 98 = _____	34 + 61 = _____
100 – 98 = _____	742 – 98 = _____	960 + 40 = _____
800 – 99 = _____	365 – 97 = _____	100 – 77 = _____
300 – 97 = _____	785 – 98 = _____	340 + 90 = _____
700 – 98 = _____	256 – 99 = _____	289 + 400 = _____
900 – 97 = _____	568 – 98 = _____	284 + 98 = _____
600 – 96 = _____	762 – 97 = _____	306 – 9 = _____
400 – 97 = _____	369 – 97 = _____	362 + 7 = _____
600 – 95 = _____	258 – 99 = _____	411 + 99 = _____
803 – 99 = _____	204 – 98 = _____	632 – 99 = _____
206 – 98 = _____	560 – 96 = _____	640 – 80 = _____
409 – 98 = _____	701 – 98 = _____	982 – 60 = _____
305 – 99 = _____	584 – 98 = _____	42 + 58 = _____
705 – 97 = _____	887 – 95 = _____	633 – 50 = _____
208 – 97 = _____	963 – 93 = _____	67 + 60 = _____
108 – 98 = _____	514 – 94 = _____	391 – 97 = _____
702 – 96 = _____	681 – 95 = _____	300 + 630 = _____

Mental Math 19	Mental Math 20
$58 + 7 =$ _____	$834 + 7 - 20 + 60 =$ _____
$85 - 32 =$ _____	$368 - 99 + 40 + 3 =$ _____
$235 + 70 =$ _____	$429 + 8 - 7 - 90 =$ _____
$984 - 200 =$ _____	$501 - 4 - 30 + 99 =$ _____
$984 - 121 =$ _____	$87 + 7 + 2 - 9 + 40 =$ _____
$587 - 98 =$ _____	$4 + 4 + 4 + 4 + 98 =$ _____
$203 + 8 =$ _____	$5 + 5 + 5 + 5 + 280 =$ _____
$698 - 3 =$ _____	$33 + 21 + 9 - 30 + 130 =$ _____
$100 - 28 =$ _____	$925 - 98 + 3 - 70 + 5 =$ _____
$98 + 654 =$ _____	$62 + 8 - 2 + 40 - 8 - 33 =$ _____
$687 - 99 =$ _____	$592 + 99 - 98 + 7 =$ _____
$414 + 8 =$ _____	$83 + 7 - 26 + 99 =$ _____
$683 - 7 =$ _____	$300 - 99 + 70 + 6 =$ _____
$610 - 90 =$ _____	$29 + 38 - 17 + 20 =$ _____
$700 + 98 =$ _____	$203 - 7 - 3 + 400 =$ _____
$250 - 98 =$ _____	$3 + 3 + 3 + 3 + 3 + 2 =$ _____
$250 + 80 =$ _____	$4 + 4 + 4 + 4 + 4 + 4 =$ _____
$100 - 62 =$ _____	$5 + 5 + 5 + 5 + 5 + 5 =$ _____
$754 - 601 =$ _____	$40 - 4 - 4 - 4 - 4 - 4 - 4 =$ _____
$986 - 123 =$ _____	$2 + 3 + 4 + 5 + 6 + 7 + 8 + 9 + 10 =$ _____

Mental Math 21	Mental Math 22	Mental Math 23
2 x 4 = _____	2 x 7 = _____	12 ÷ 4 = _____
4 x 9 = _____	4 x 8 = _____	4 ÷ 4 = _____
6 x 4 = _____	9 x 3 = _____	36 ÷ 4 = _____
4 x 7= _____	4 x 5 = _____	32 ÷ 4 = _____
4 x 6 = _____	2 x 6 = _____	24 ÷ 4 = _____
4 x 3 = _____	8 x 2 = _____	40 ÷ 4 = _____
7 x 4 = _____	4 x 6 = _____	20 ÷ 4 = _____
4 x 10 = _____	9 x 4 = _____	4 ÷ 4 = _____
4 x 5 = _____	5 x 3 = _____	16 ÷ 4 = _____
8 x 4 = _____	2 x 9 = _____	8 ÷ 4 = _____
5 x 4 = _____	4 x 4 = _____	16 ÷ 4 = _____
4 x 9 = _____	3 x 8 = _____	28 ÷ 4 = _____
4 x 4 = _____	6 x 3 = _____	32 ÷ 4 = _____
7 x 4 = _____	2 x 5 = _____	8 ÷ 4 = _____
4 x 8 = _____	3 x 2 = _____	12 ÷ 4 = _____
4 x 6 = _____	3 x 7 = _____	20 ÷ 4 = _____
4 x 10 = _____	4 x 3 = _____	28 ÷ 4 = _____
4 x 2 = _____	3 x 3 = _____	16 ÷ 4 = _____
8 x 4 = _____	7 x 4 = _____	24 ÷ 4 = _____
3 x 4 = _____	2 x 4 = _____	40 ÷ 4 = _____

Mental Math 24	Mental Math 25	Mental Math 26
12 ÷ 3 = _____	4 x 4 = _____	1 x 5 = _____
36 ÷ 4 = _____	36 ÷ 4 = _____	6 x 5 = _____
18 ÷ 2 = _____	5 x 4 = _____	4 x 5 = _____
24 ÷ 4 = _____	28 ÷ 4 = _____	5 x 7 = _____
15 ÷ 3 = _____	30 ÷ 3 = _____	5 x 2 = _____
32 ÷ 4 = _____	4 x 7 = _____	3 x 5 = _____
18 ÷ 3 = _____	9 x 4 = _____	5 x 10 = _____
16 ÷ 2 = _____	24 ÷ 4 = _____	8 x 5 = _____
40 ÷ 4 = _____	4 x 6 = _____	5 x 5 = _____
20 ÷ 4 = _____	20 ÷ 4 = _____	10 x 5 = _____
21 ÷ 3 = _____	3 x 8 = _____	5 x 3 = _____
28 ÷ 4 = _____	18 ÷ 3 = _____	5 x 6 = _____
24 ÷ 3 = _____	21 ÷ 3 = _____	5 x 9 = _____
27 ÷ 3 = _____	4 x 5 = _____	2 x 5 = _____
24 ÷ 4 = _____	8 ÷ 4 = _____	5 x 5 = _____
8 ÷ 4 = _____	32 ÷ 4 = _____	7 x 5 = _____
16 ÷ 4 = _____	10 x 4 = _____	5 x 1 = _____
20 ÷ 2 = _____	12 ÷ 4 = _____	5 x 8 = _____
16 ÷ 4 = _____	40 ÷ 4 = _____	9 x 5 = _____
12 ÷ 2 = _____	7 x 3 = _____	5 x 4 = _____

Mental Math 27	Mental Math 28	Mental Math 29
$3 \times 6 =$ _____	$35 \div 5 =$ _____	$30 \div 5 =$ _____
$5 \times 2 =$ _____	$27 \div 3 =$ _____	$35 \div 5 =$ _____
$7 \times 4 =$ _____	$5 \div 5 =$ _____	$10 \div 2 =$ _____
$3 \times 5 =$ _____	$16 \div 2 =$ _____	$40 \div 5 =$ _____
$3 \times 9 =$ _____	$15 \div 5 =$ _____	$32 \div 4 =$ _____
$5 \times 4 =$ _____	$36 \div 4 =$ _____	$50 \div 5 =$ _____
$3 \times 4 =$ _____	$25 \div 5 =$ _____	$20 \div 2 =$ _____
$5 \times 5 =$ _____	$9 \div 3 =$ _____	$25 \div 5 =$ _____
$4 \times 4 =$ _____	$30 \div 5 =$ _____	$10 \div 5 =$ _____
$6 \times 5 =$ _____	$18 \div 2 =$ _____	$15 \div 3 =$ _____
$5 \times 7 =$ _____	$45 \div 5 =$ _____	$45 \div 5 =$ _____
$4 \times 8 =$ _____	$24 \div 3 =$ _____	$20 \div 5 =$ _____
$9 \times 5 =$ _____	$50 \div 5 =$ _____	$24 \div 3 =$ _____
$4 \times 9 =$ _____	$16 \div 4 =$ _____	$40 \div 4 =$ _____
$10 \times 5 =$ _____	$30 \div 3 =$ _____	$15 \div 5 =$ _____
$7 \times 3 =$ _____	$12 \div 2 =$ _____	$18 \div 3 =$ _____
$4 \times 2 =$ _____	$40 \div 5 =$ _____	$5 \div 5 =$ _____
$5 \times 8 =$ _____	$10 \div 5 =$ _____	$12 \div 3 =$ _____
$6 \times 4 =$ _____	$28 \div 4 =$ _____	$21 \div 3 =$ _____
$10 \times 4 =$ _____	$20 \div 5 =$ _____	$12 \div 4 =$ _____

Mental Math 30	Mental Math 31	Mental Math 32
10 x 7 = _____	20 ÷ 10 = _____	15 ÷ 5 = _____
5 x 8 = _____	100 ÷ 10 = _____	28 ÷ 4 = _____
10 x 2 = _____	30 ÷ 3 = _____	5 x 9 = _____
3 x 9 = _____	30 ÷ 10 = _____	4 x 5 = _____
4 x 10 = _____	10 ÷ 5 = _____	20 ÷ 4 = _____
5 x 10 = _____	50 ÷ 10 = _____	32 ÷ 4 = _____
8 x 4 = _____	10 ÷ 2 = _____	4 x 4 = _____
10 x 6 = _____	70 ÷ 10 = _____	25 ÷ 5 = _____
9 x 5 = _____	30 ÷ 5 = _____	6 x 5 = _____
8 x 10 = _____	60 ÷ 10 = _____	4 x 2 = _____
7 x 2 = _____	50 ÷ 5 = _____	80 ÷ 10 = _____
10 x 10 = _____	10 ÷ 10 = _____	4 x 6 = _____
9 x 4 = _____	40 ÷ 5 = _____	10 x 7 = _____
8 x 3 = _____	40 ÷ 4 = _____	20 ÷ 5 = _____
10 x 1 = _____	90 ÷ 10 = _____	10 x 6 = _____
7 x 3 = _____	20 ÷ 5 = _____	4 x 8 = _____
3 x 10 = _____	40 ÷ 10 = _____	10 x 5 = _____
9 x 10 = _____	20 ÷ 4 = _____	8 x 5 = _____
4 x 7 = _____	80 ÷ 10 = _____	24 ÷ 3 = _____
5 x 7 = _____	20 ÷ 2 = _____	3 x 5 = _____

Mental Math 33	Mental Math 34	Mental Math 35
2 x 3 = _____	100 – 29 = _____	100 – 75 = _____
12 ÷ 4 = _____	876 + 7 = _____	50 ÷ 5 = _____
80 ÷ 10 = _____	24 ÷ 4 = _____	73 + 8 = _____
8 ÷ 2 = _____	68 + 90 = _____	68 + 80 = _____
2 x 5 = _____	361 + 50 = _____	216 + 9 = _____
100 ÷ 10 = _____	637 + 41 = _____	167 + 70 = _____
24 ÷ 3 = _____	5 x 9 = _____	37 + 31 = _____
21 ÷ 3 = _____	98 + 238 = _____	23 – 9 = _____
24 ÷ 4 = _____	42 – 6 = _____	420 – 8 = _____
3 x 3 = _____	954 – 7 = _____	920 – 50 = _____
3 x 6 = _____	123 – 90 = _____	10 x 9 = _____
15 ÷ 3 = _____	47 – 16 = _____	57 – 35 = _____
8 x 10 = _____	762 – 98 = _____	584 – 98 = _____
27 ÷ 3 = _____	800 – 99 = _____	986 – 400 = _____
32 ÷ 4 = _____	5 x 4 = _____	95 + 198 = _____
4 x 3 = _____	35 ÷ 5 = _____	36 ÷ 4 = _____
24 ÷ 3 = _____	59 + 5 = _____	7 x 5 = _____
4 x 7 = _____	7 x 4 = _____	34 + 61 = _____
10 x 10 = _____	100 ÷ 10 = _____	232 + 600 = _____
6 x 10 = _____	683 – 7 = _____	4 x 8 = _____

Mental Math 36	Mental Math 37
75¢ = _____ quarters	$1 – 75¢ = _____ ¢
60¢ = _____ dimes	$1 – 50¢ = _____ ¢
55¢ = _____ nickels	$1 – 35¢ = _____ ¢
50¢ = _____ quarters	$1 – 60¢ = _____ ¢
$1 = _____ quarters	$1 – 25¢ = _____ ¢
$6 = _____ quarters	$1 – 95¢ = _____ ¢
$2 = _____ nickels	85¢ + _____ ¢ = $1
$3 = _____ pennies	45¢ + _____ ¢ = $1
$5 = _____ dimes	10¢ + _____ ¢ = $1
$3.80 = _____ dimes	_____ ¢ + 30¢ = $1
$4.25 = _____ quarters	11¢ + _____ ¢ = $1
$1.20 = _____ nickels	_____ ¢ + 7¢ = $1
$8.50 = _____ quarters	$1 – 32¢ = _____ ¢
$12.50 = _____ dimes	49¢ + _____ ¢ = $1
$20 = _____ 5-dollar bills	94¢ + _____ ¢ = $1
$10 = _____ 1-dollar bills	$1 – 56¢ = _____ ¢
$40 = _____ 20-dollar bills	_____ ¢ + 72¢ = $1
$100 = _____ 10-dollar bills	$1 – 13¢ = _____ ¢
$45 = _____ 5-dollar bills	_____ ¢ + 64¢ = $1
$30 = _____ 5-dollar bills	_____ ¢ + 28¢ = $1

Mental Math 38	Mental Math 39
$10 − $0.85 = $_____	$2.65 + $3 = $_____
$10 − $0.40 = $_____	$1.25 + 75¢ = $_____
$10 − $2.30 =$_____	$3.15 + 55¢ = $_____
$3.60 + $_____ = $10	$5.32 + $2 = $_____
$8.35 + $_____= $10	$8.35 + 25¢ = $_____
$2.70 + $_____= $10	$1.32 + 27¢ = $_____
$10 − $3.65 = $_____	$2.65 + $4 = $_____
$10 − $4.20 = $_____	$2.41 + 32¢ = $_____
$10 − $2.05 = $_____	$9.20 + 62¢ = $_____
$7.10 + $_____= $10	$3.20 + $4.40 = $_____
$8.35 + $_____= $10	$3.35 + $2.15 = $_____
$_____ + $4.15 = $10	$6.40 + $0.15 = $_____
$10 − $5.75 = $_____	$2.25 + $4.35 = $_____
$4.93 + $_____= $10	$2.40 + $3.18 = $_____
$9.36 + $_____= $10	$6.24 + $1.07 = $_____
$10 − $5.61 = $_____	$2.41 + $6.25 = $_____
$_____+ $7.28 = $10	$4.35 + $4.49 = $_____
$10 − $0.12 = $_____	$2.05 + $6.95 = $_____
$_____ + $8.44 = $10	$0.30 + $6.70 = $_____
$_____ + $1.27 = $10	$2.45 + $5.55 = $_____

Mental Math 40	Mental Math 41
$6.40 + $0.99 = $_____	$8.60 – $4 = $_____
$2.25 + $0.98 = $_____	68¢ – 21¢ = $_____
$2.43 + $3.98 = $_____	$9.35 – $7 = $_____
$6.24 + $1.97 = $_____	$4 – 75¢ = $_____
$2.20 + $6.95 = $_____	$8 – 25¢ = $_____
$4.35 + $4.99 = $_____	$5.15 – $3 = $_____
$2.05 + $6.95 = $_____	$5 – 40¢ = $_____
$3.45 + $2.96 = $_____	$3.83 – 25¢ = $_____
$1.25 + $0.97 = $_____	$3 – 29¢ = $_____
$4.65 + $3.98 = $_____	$4.90 – $1.80 = $_____
$6.97 + $1.24 = $_____	$8.85 – $6.05 = $_____
$7.30 + $0.95 = $_____	$3.45 – $1.25 = $_____
$2.96 + $1.37 = $_____	$9.95 – $4.30 = $_____
$2.05 + $6.95 = $_____	$7.70 – $3.35 = $_____
$1.07 + $3.97 = $_____	$6.65 – $2.45 = $_____
$3.96 + $2.55 = $_____	$3.85 – $2.15 = $_____
$4.95 + $2.24 = $_____	$42 – $8 = $_____
$0.20 + $1.96 = $_____	$95 – $7 = $_____
$2.95 + $3.37 = $_____	$387 – $98 = $_____
$1.95 + $6.99 = $_____	$600 – $42 = $_____

Mental Math 42	Mental Math 43
$3 – $0.99 = $_____	$6.35 – $1.97 = $_____
$5 – $0.96 = $_____	$348 + $98 = $_____
$8 – $0.97 = $_____	$329 – $98 = $_____
$4.35 – $0.98 = $_____	$3.05 + $6.95 = $_____
$7.25 – $0.96 = $_____	$10.00 – $6.40 = $_____
$3.65 – $0.95 = $_____	$4.65 + $3.99 = $_____
$5.80 – $3.96 = $_____	$1.00 – $0.35 = $_____
$9.05 – $6.97 = $_____	$1.00 – $0.85 = $_____
$6.70 – $2.95 = $_____	$10.00 – $2.75 =$_____
$2.85 – $0.97 = $_____	$9.27 + $0.66 = $_____
$7.69 – $3.95 = $_____	$3.83 – $2.25 = $_____
$6.25 – $0.95 = $_____	$3.21 + $4.72 = $_____
$4.85 – $3.96 = $_____	$3.35 + $2.99 = $_____
$9.32 – $6.95 = $_____	$5.00 – $0.45 = $_____
$6.29 – $2.95 = $_____	$3.00 – $0.35 = $_____
$5.82 – $0.97 = $_____	$4.05 – $1.75 = $_____
$6.67 – $5.95 = $_____	$2.95 + $3.50 = $_____
$8.35 – $2.96 = $_____	$8.75 – $6.50 = $_____
$1.98 – $0.99 = $_____	$1.25 + $0.90 = $_____
$9.55 – $8.95 = $_____	$3.65 + $5.75 = $_____

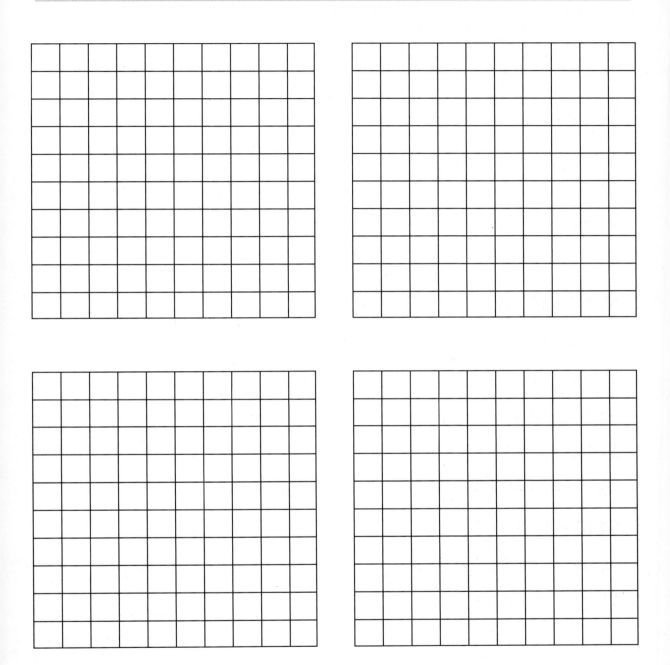

1	2	3	4	5	6	7	8	9	10
11	12	13	14	15	16	17	18	19	20
21	22	23	24	25	26	27	28	29	30
31	32	33	34	35	36	37	38	39	40

1	2	3	4	5	6	7	8	9	10
11	12	13	14	15	16	17	18	19	20
21	22	23	24	25	26	27	28	29	30
31	32	33	34	35	36	37	38	39	40
41	42	43	44	45	46	47	48	49	50

$4 \times 1 =$ _____ $1 \times 4 =$ _____

$4 \times 2 =$ _____ $2 \times 4 =$ _____

$4 \times 3 =$ _____ $3 \times 4 =$ _____

$4 \times 4 =$ _____ $4 \times 4 =$ _____

$4 \times 5 =$ _____ $5 \times 4 =$ _____

$4 \times 6 =$ _____ $6 \times 4 =$ _____

$4 \times 7 =$ _____ $7 \times 4 =$ _____

$4 \times 8 =$ _____ $8 \times 4 =$ _____

$4 \times 9 =$ _____ $9 \times 4 =$ _____

$4 \times 10 =$ _____ $10 \times 4 =$ _____

x	1	2	3	4	5	6	7	8	9	10
2										
3										
4										

x	5	3	6	1	4	10	2	9	8	7
3										
4										
2										

x	4	10	9	6	2	7	3	8	1	5
4										
2										
3										

_____ x 4 = 4   4 ÷ 4 = _____

_____ x 4 = 8   8 ÷ 4 = _____

_____ x 4 = 12   12 ÷ 4 = _____

_____ x 4 = 16   16 ÷ 4 = _____

_____ x 4 = 20   20 ÷ 4 = _____

_____ x 4 = 24   24 ÷ 4 = _____

_____ x 4 = 28   28 ÷ 4 = _____

_____ x 4 = 32   32 ÷ 4 = _____

_____ x 4 = 36   36 ÷ 4 = _____

_____ x 4 = 40   40 ÷ 4 = _____

5 x 1 = _____   1 x 5 = _____

5 x 2 = _____   2 x 5 = _____

5 x 3 = _____   3 x 5 = _____

5 x 4 = _____   4 x 5 = _____

5 x 5 = _____   5 x 5 = _____

5 x 6 = _____   6 x 5 = _____

5 x 7 = _____   7 x 5 = _____

5 x 8 = _____   8 x 5 = _____

5 x 9 = _____   9 x 5 = _____

5 x 10 = _____   10 x 5 = _____

x	1	2	3	4	5	6	7	8	9	10
3										
4										
5										

x	5	3	6	1	4	10	2	9	8	7
4										
5										
2										

x	4	10	9	6	2	7	3	8	1	5
5										
4										
3										

_____ x 5 = 5 $5 \div 5 =$ _____

_____ x 5 = 10 $10 \div 5 =$ _____

_____ x 5 = 15 $15 \div 5 =$ _____

_____ x 5 = 20 $20 \div 5 =$ _____

_____ x 5 = 25 $25 \div 5 =$ _____

_____ x 5 = 30 $30 \div 5 =$ _____

_____ x 5 = 35 $35 \div 5 =$ _____

_____ x 5 = 40 $40 \div 5 =$ _____

_____ x 5 = 45 $45 \div 5 =$ _____

_____ x 5 = 50 $50 \div 5 =$ _____

10 x 1 = _____   1 x 10 = _____

10 x 2 = _____   2 x 10 = _____

10 x 3 = _____   3 x 10 = _____

10 x 4 = _____   4 x 10 = _____

10 x 5 = _____   5 x 10 = _____

10 x 6 = _____   6 x 10 = _____

10 x 7 = _____   7 x 10 = _____

10 x 8 = _____   8 x 10 = _____

10 x 9 = _____   9 x 10 = _____

10 x 10 = _____   10 x 10 = _____

_____ x 10 = 10 $10 \div 10 =$ _____

_____ x 10 = 20 $20 \div 10 =$ _____

_____ x 10 = 30 $30 \div 10 =$ _____

_____ x 10 = 40 $40 \div 10 =$ _____

_____ x 10 = 50 $50 \div 10 =$ _____

_____ x 10 = 60 $60 \div 10 =$ _____

_____ x 10 = 70 $70 \div 10 =$ _____

_____ x 10 = 80 $80 \div 10 =$ _____

_____ x 10 = 90 $90 \div 10 =$ _____

_____ x 10 = 100 $100 \div 10 =$ _____

1											

| $\frac{1}{2}$ | | | | | | $\frac{1}{2}$ | | | | | |

| $\frac{1}{3}$ | | | | $\frac{1}{3}$ | | | | $\frac{1}{3}$ | | | |

| $\frac{1}{4}$ | | | $\frac{1}{4}$ | | | $\frac{1}{4}$ | | | $\frac{1}{4}$ | | |

| $\frac{1}{5}$ | | $\frac{1}{5}$ | | $\frac{1}{5}$ | | $\frac{1}{5}$ | | $\frac{1}{5}$ | | | |

| $\frac{1}{6}$ | | $\frac{1}{6}$ | | $\frac{1}{6}$ | | $\frac{1}{6}$ | | $\frac{1}{6}$ | | $\frac{1}{6}$ | |

| $\frac{1}{7}$ | $\frac{1}{7}$ | $\frac{1}{7}$ | $\frac{1}{7}$ | $\frac{1}{7}$ | $\frac{1}{7}$ | $\frac{1}{7}$ | | | | | |

| $\frac{1}{8}$ | $\frac{1}{8}$ | $\frac{1}{8}$ | $\frac{1}{8}$ | $\frac{1}{8}$ | $\frac{1}{8}$ | $\frac{1}{8}$ | $\frac{1}{8}$ | | | | |

| $\frac{1}{9}$ | $\frac{1}{9}$ | $\frac{1}{9}$ | $\frac{1}{9}$ | $\frac{1}{9}$ | $\frac{1}{9}$ | $\frac{1}{9}$ | $\frac{1}{9}$ | $\frac{1}{9}$ | | | |

| $\frac{1}{10}$ | $\frac{1}{10}$ | $\frac{1}{10}$ | $\frac{1}{10}$ | $\frac{1}{10}$ | $\frac{1}{10}$ | $\frac{1}{10}$ | $\frac{1}{10}$ | $\frac{1}{10}$ | $\frac{1}{10}$ | | |

| $\frac{1}{11}$ | $\frac{1}{11}$ | $\frac{1}{11}$ | $\frac{1}{11}$ | $\frac{1}{11}$ | $\frac{1}{11}$ | $\frac{1}{11}$ | $\frac{1}{11}$ | $\frac{1}{11}$ | $\frac{1}{11}$ | $\frac{1}{11}$ | |

| $\frac{1}{12}$ | $\frac{1}{12}$ | $\frac{1}{12}$ | $\frac{1}{12}$ | $\frac{1}{12}$ | $\frac{1}{12}$ | $\frac{1}{12}$ | $\frac{1}{12}$ | $\frac{1}{12}$ | $\frac{1}{12}$ | $\frac{1}{12}$ | $\frac{1}{12}$ |

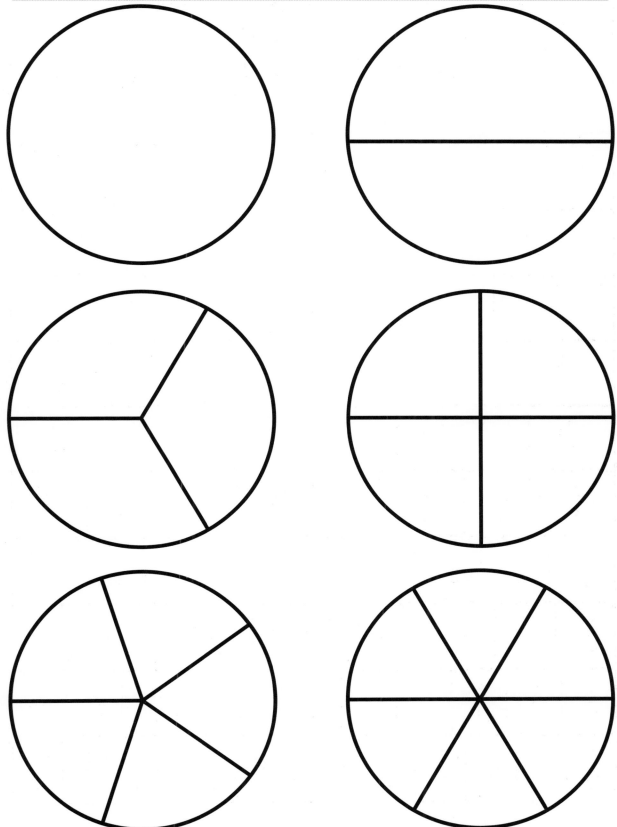

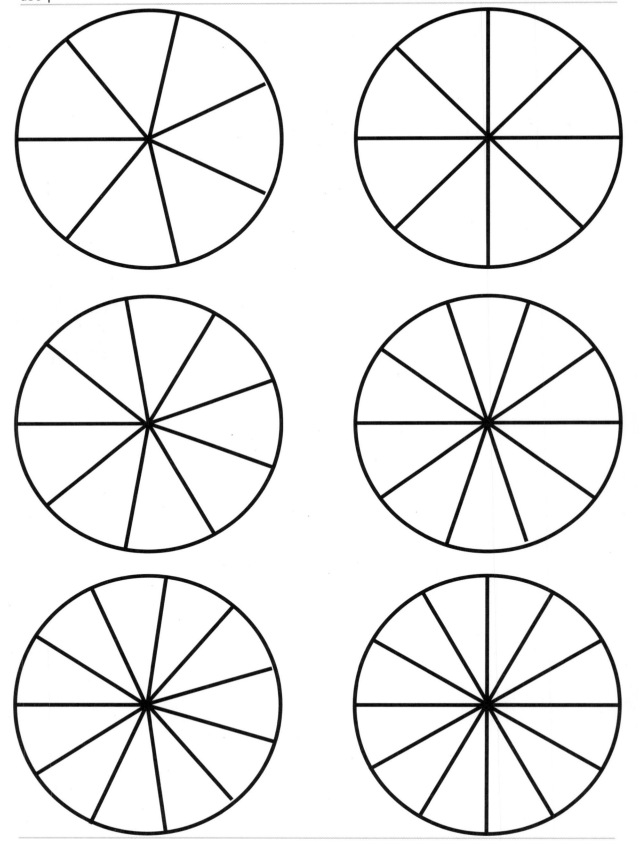

What fraction of the shapes are circles?

What fraction of the shapes are striped?

What fraction of the shapes are dotted?

What fraction of the shapes have 4 sides?

What fraction of the shapes are black?

Fruit		
Mango		
Pear		
Apple		
Orange		

Each ◯ stands for

Decide on a symbol and how much the symbol should stand for. Then Complete the graph below to show how much money each student has.

Name	Amount of money in dollars
Peter	18
Jerry	24
Paula	15
Mike	12
Mary	$30

Amount of Money
Peter
Jerry
Paula
Mike
Mary

Each _____ stands for _____

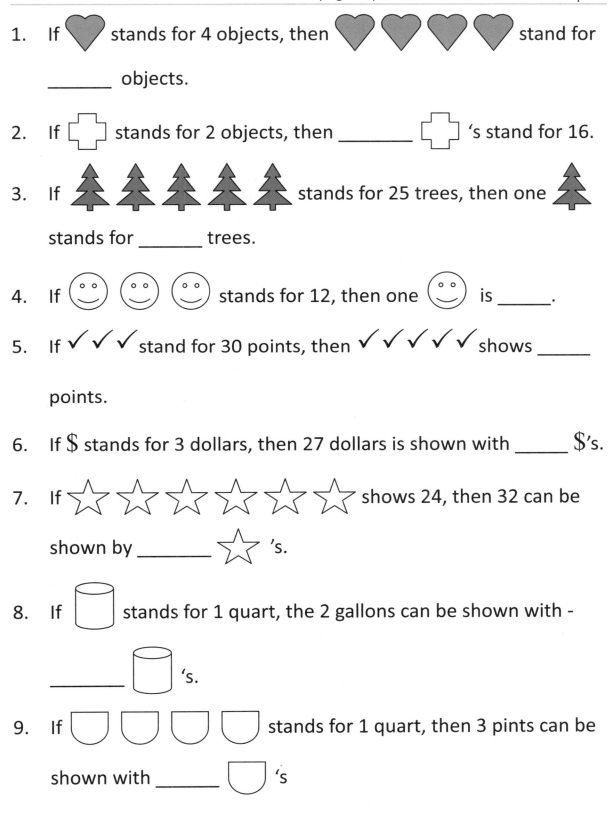

1. If ♥ stands for 4 objects, then ♥ ♥ ♥ ♥ stand for

   _____ objects.

2. If ✚ stands for 2 objects, then _____ ✚'s stand for 16.

3. If 🌲 🌲 🌲 🌲 🌲 stands for 25 trees, then one 🌲

   stands for _____ trees.

4. If ☺ ☺ ☺ stands for 12, then one ☺ is _____.

5. If ✓ ✓ ✓ stand for 30 points, then ✓ ✓ ✓ ✓ ✓ shows _____

   points.

6. If $ stands for 3 dollars, then 27 dollars is shown with _____ $'s.

7. If ☆ ☆ ☆ ☆ ☆ ☆ shows 24, then 32 can be

   shown by _____ ☆'s.

8. If ⬭ stands for 1 quart, the 2 gallons can be shown with -

   _____ ⬭'s.

9. If ⋃ ⋃ ⋃ ⋃ stands for 1 quart, then 3 pints can be

   shown with _____ ⋃'s

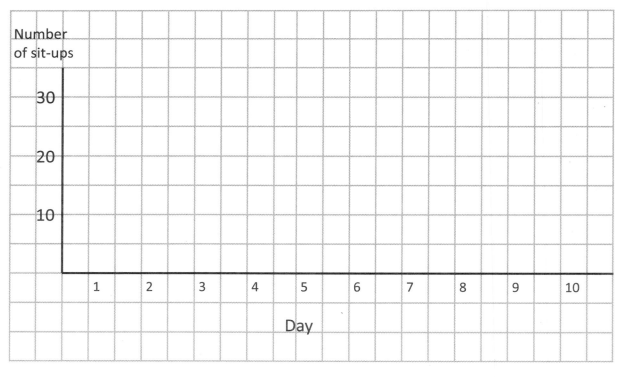

Number of books read

80
70
60
50
40
30
20
10

Selena   David   Maria   Meiling   Baljit

Number of sit-ups

30

20

10

1   2   3   4   5   6   7   8   9   10

Day

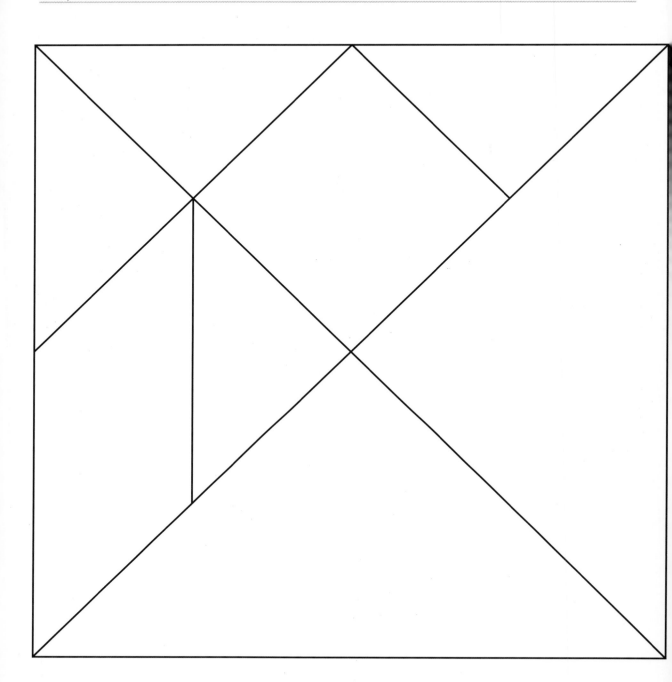

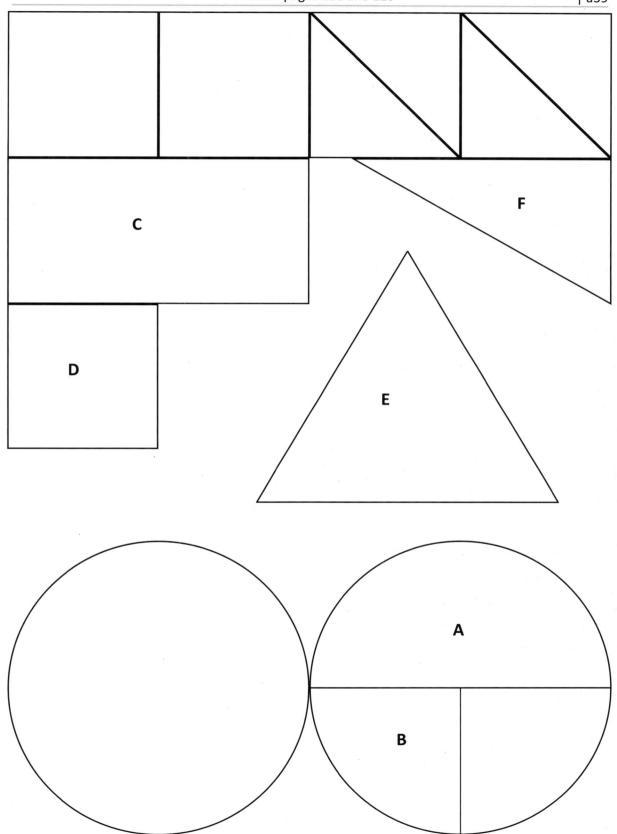

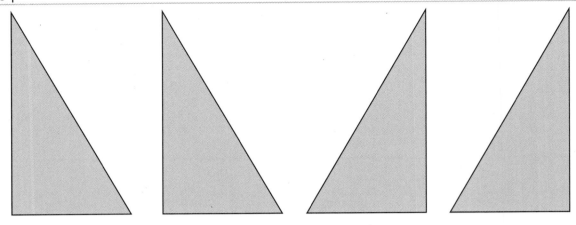

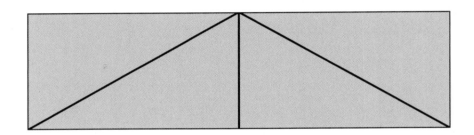

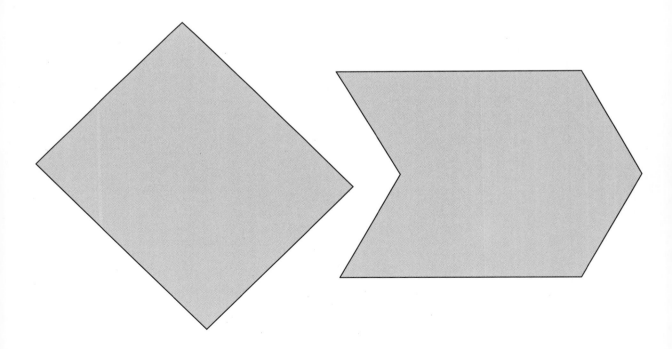

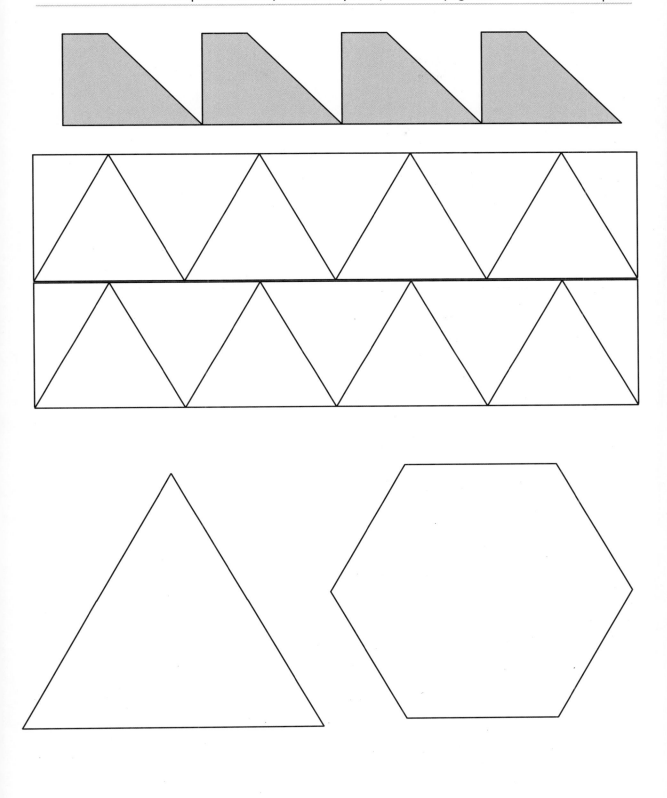

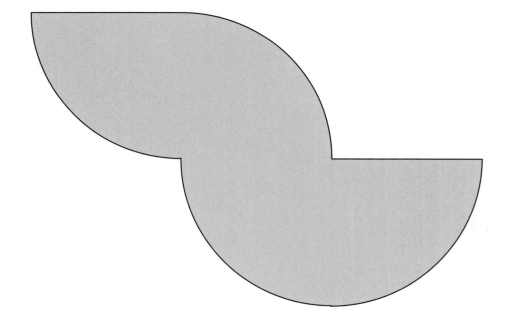

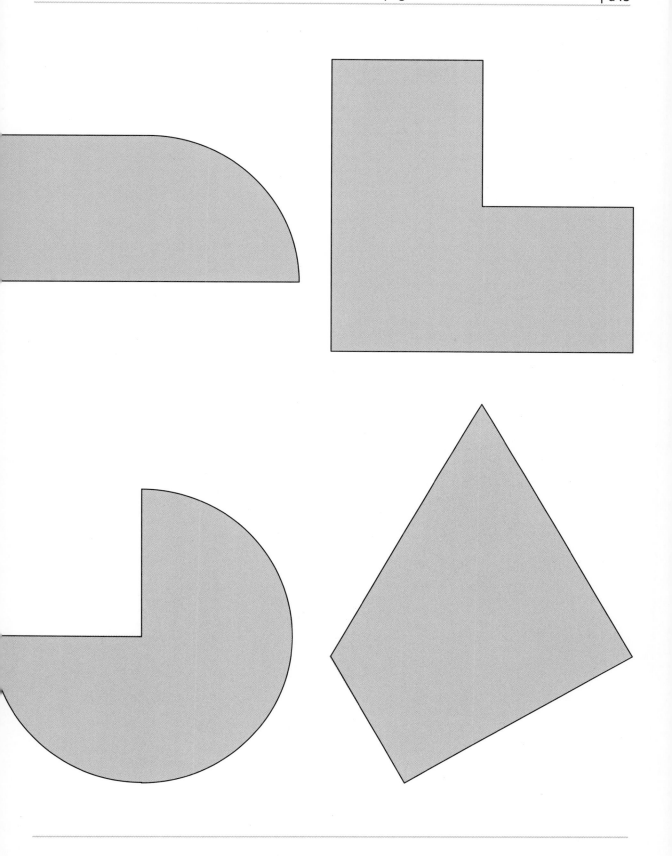